THE DEMON
IN THE AETHER

THE DEMON
IN THE AETHER
The Story of James Clerk Maxwell

Martin Goldman

Paul Harris Publishing

Edinburgh
In association with Adam Hilger Ltd

First published 1983 by
Paul Harris Publishing
40 York Place
Edinburgh

In association with Adam Hilger Ltd
Techno House, Bristol, BS1 6NX

ISBN 0 86228 026 5

British Library Cataloguing in Publication Data

Goldman, Martin
   The demon in the aether: the story of James
   Clerk Maxwell.
   1. Clerk Maxwell, James 2. Physics—Scotland
  —Biography
   I. Title
   530'.092'4   QC16.M4

   ISBN 0-86228-026-5

*The publishers gratefully acknowledge the financial assistance
of the Scottish Arts Council in the publication of this volume.*

*Printed and bound in Great Britain by
Spectrum Printing Company, Edinburgh*

CONTENTS

ILLUSTRATIONS

1. Map of the birthplaces of 19th century scientists.

2. Maxwell as a boy from one of his cousin Jemima's watercolours. *(Cavendish Laboratory, University of Cambridge)*.

3. Maxwell: the young scientist at Cambridge. He is holding the "improved" colour top. *(Cambridge University Library)*.

4. Professor Maxwell. *(Peterhouse College, Cambridge)*.

5. Professor Maxwell from a drawing made when he was the Cavendish Professor. The strain of establishing the laboratory (amongst other things) seems to have turned his hair grey. *(Cavendish Laboratory)*.

6. Professor and Mrs Maxwell plus a shaggy dog, posed in front of a painted backdrop to simulate a suitably romantic Highland atmosphere. This is the only photograph for which Maxwell kept on his spectacles. *(Cavendish Laboratory)*.

7. Maxwell's dynamical top, with 9 adjusting screws and a colour disc on top. *(Cavendish Laboratory)*.

8. The model made by Ramage in Aberdeen to demonstrate Saturn's rings. A handle is turned at the rear and the little ivory balls at the front can wiggle in two of the different wave-modes of a single ring. *(Cavendish Laboratory)*.

9. Apparatus to measure the viscosity of gases. Its heat jacket has been lost, but at least that allows one to see the fixed and rotating vanes inside. *(Cavendish Laboratory)*.

10. The balance arm wired up to compare electromagnetic against electrostatic units of charge, which ratio, at Maxwell's hands, gave the speed of light. *(Cavendish Laboratory)*.

11. Apparatus with which Maxwell and MacAlister repeated and improved Cavendish's experiment to test the validity of the inverse square law of electrostatic repulsion. They found it true to one part in 21,600. *(Cavendish Laboratory)*.

PREFACE

I have received help from many people, too many to name them all individually, but I would like to thank the following for advice during the writing of the book: C.A. Boardman, Dr Brian Bowers, Sam Callander, Professor T.G. Cowling, Professor Cyril Domb, Dr Nicholas Fisher, Professor R.V. Jones, Professor Brian Pippard, Dr R. Porter, David Standley, Brigadier John Wedderburn-Maxwell. Any misapprehensions under which I laboured, are, of course, my fault. I would also like to thank the library staff of the University of Glasgow, the Syndics of Cambridge University Library, the Royal Society, the National Library of Scotland and the University of St Andrews for access to their collections; my publisher Paul Harris and editor Trevor Royle for their essential initial encouragement of a commission and for the subsequent care and attention they have taken; my mother and brother for their encouragement; my daughter Rebecca for her forebearance and, of course, my wife, Frances, for all these things.

During the writing of this book I have come to appreciate James Clerk Maxwell more and more, both as scientist and man. I hope that a little of his remarkable quality has emerged.

Martin Goldman
Glasgow, April 1983

DEDICATION

For my father Kazimierz Goldman, who introduced me to
Maxwell's genius

CHAPTER 1

James Clerk Maxwell is one of the greatest figures in the history of science. Making lists of all-time-greats is an occupation suitable only for schoolboys, selecting teams for epic, imaginary games of cricket on rainy summer afternoons (Stonewall Jackson and the Duke of Wellington would have made a dynamic opening pair). But in physics a clear division does emerge: Archimedes, Newton, Maxwell and Einstein are in a different league from the rest. The other three are all familiar names. Archimedes was the very first streaker, Newton was bruised by falling fruit (he could not catch, was obviously a terrible fielder and so excludes himself from selection among the cricketing superstars), and Einstein had a shock of white hair, a fiddle and no socks. Popular legend, however, has never managed to associate Maxwell with any such quaint trait, to reduce him to common mortality, and so, inevitably, he has been ignored.

Yet his achievements were extraordinary. He completely revolutionised two fields of physics, statistical mechanics and electromagnetism. In statistical mechanics, he took the idea that a gas was made up of molecules, which by bouncing against the walls of their container effectively exerted pressure, and worked out how these molecules would jostle against each other, some fast, some slow. To calculate the motion of each molecule would be impossible, but he was able to show that the gas would settle down to a determinate statistical scatter of velocities—the Maxwell distribution. He thus introduced the idea of probability into physics, and with it was able to calculate many of the properties of gases. For example, he predicted that the viscosity of a gas—how treacly it is, how it resists things slipping through it—should be independent of pressure. This was a startling result, but Maxwell then proceeded himself to perform the experiment that proved it correct.

In electromagnetism, there was a large body of experimental data, and several independent theories to explain its various parts. Maxwell took the unifying notion of 'lines of force' which Faraday had invented, and welded the separate topics of electricity and magnetism into a unified whole, which could be expressed in just four simple mathematical formulae—Maxwell's equations. By combining them he predicted that it should be possible to transmit an electromagnetic wave that would beam off into space. When he worked out its velocity, it transpired that it should be the same as that of light: light was merely one manifestation of his waves. Optics was now no longer a separate branch of physics, but a subdivision of electromagnetism. Though its existence was not proved till after his death, electromagnetic radiation at wavelengths longer than light has made possible radio and television in the present century, and at shorter wavelengths has given us X-rays. The new theoretical technique that Maxwell used in this work: taking 'lines of force' as serious mathematical entities, field theory, has been central to the development of twentieth century physics.

In optics, Maxwell did fundamental work on the physics and physiology of colour vision. He helped prove that the normal eye has three sorts of receptor, one sensitive to red light, another to blue and the third to green. He showed colour blindness to be the result of the absence of one of those receptors, and investigated other defects of vision. A particular effect involving a spot seen by most people when looking at blue light is called the Maxwell spot. Photography was then rapidly becoming a popular Victorian pastime, and Maxwell contributed to it by taking the first ever colour photograph.

In thermodynamics, Maxwell wrote a standard text-book, and almost without being aware of the fact produced a useful set of relations known, confusingly, again as Maxwell's equations. In astronomy, he looked at the structure of the rings around Saturn, and showed it was impossible for them to be either solid or fluid. They had to be made up of lots of solid particles; flights of brickbats, he called them. Maxwell also made important discoveries in the mathematical subject of topology, wrote the founding paper in the field of cybernetics, invented reciprocal diagrams, which are useful in engineering, looked at the theory of optical instruments, the theory of strain-induced double refraction in solids, and, among other things, how a sheet of paper tumbles to the floor.

That is a phenomenal body of work of the very highest quality; Professor C.A. Coulson once said: 'There is scarcely a single topic that he touched upon which he did not change almost beyond recognition'.[1] And yet Maxwell died when he was only forty-eight years old.

As for the man himself, the best description that we have comes from an unnamed source quoted in Lewis Campbell's official biography, someone who is cryptically referred to as having first seen Maxwell in 1866, when he would have been thirty-five:

> A man of middle height, with frame strongly knit, and a certain spring and elasticity in his gait; dressed for comfortable ease rather than elegance; a face expressive at once of sagacity and good humour, but overlaid with a deep shade of thoughfulness; features boldly but pleasingly marked; eyes dark and glowing; hair and beard perfectly black, and forming a strong contrast to the pallor of his complexion. . . . He might have been taken, by a careless observer, for a country gentleman, or rather, to be more accurate, for a north country laird. A keener eye would have seen, however, that the man must be a student of some sort, and one of more than ordinary intelligence.
>
> The picture of Maxwell, as he appeared in 1866, became afterwards perfectly familiar to residents in Cambridge. They will remember his thoughtful face as he walked in the street, revolving some of the many problems that engaged him, Toby lagging behind, till his master would suddenly turn, as if starting from a

reverie, and begin calling the dog. . . . He had a strong sense of humour, and a keen relish for witty or jocose repartee, but rarely betrayed enjoyment by outright laughter. The outward sign and conspicuous manifestation of his enjoyment was a peculiar twinkle and brightness of the eyes. There was, indeed, nothing explosive in his mental composition, and as his mirth was never boisterous, so neither was he fretful or irascible. Of a serenely placid temper, genial and temperate in his enjoyments, and infinitely patient when others would have been vexed or annoyed, he at all times opposed a solid calm of nature to the vicissitudes of life.

In performing his private experiments at the laboratory, Maxwell was very neat-handed and expeditious. When working thus, or when thinking out a problem, he had a habit of whistling, not loudly, but in a half-subdued manner, no particular tune discernible, but a sort of running accompaniment to his inward thoughts. . . . He could carry the full strength of his mental faculties rapidly from one subject to another, and could pursue his studies under distractions which most students would find intolerable, such as a loud conversation in the room where he was at work. On these occasions he used, in a manner, to take his dog into his confidence, and would say softly, "Tobi, Tobi," at intervals.

. . . His aquaintance with the literature of his own country, and especially with English poetry, was remarkable alike for its extent, its exactness, and the wide range of his sympathies. His critical taste, founded as it was on his native sagacity, and a keen appreciation of literary beauty, was so true and dis-criminating that his judgment was, in such matters, quite as valuable as on mathematical writings. . . . As he read with great rapidity, and had a retentive memory, his mind was stored with many a choice fragment which had caught his fancy. He was fond of reading aloud at home from his favourite authors, particularly from Shakespeare, and of repeating such passages as gave him the greatest pleasure.[2]

For the purposes of biography there is another fascinating aspect to Maxwell: he thought deeply about the philosophy and psychology of science, and has left us, albeit mostly unwittingly, a good deal of information about how his particular genius functioned.

There is always a special difficulty in trying to understand genius. Of course, to understand the *detailed* workings of *anyone's* mind is impossible—the brain is too complex—but with 'normal' people, it is at least conceivable that, with enough background information, the major decisions of their lives should be intelligible to another 'normal' person. Not so with genius; there must be something essentially unintelligible about them, the separate segments of their personalities must add to a whole greater than our comprehension—otherwise the recipe for genius would have been analysed, and production lines set up to churn them out like motor cars. Maxwell himself would probably have agreed with this. In a penportrait of his friend Faraday for the journal *Nature* he wrote in 1873:

Every great man of the first rank is unique. Each has his own office and his own place in the historic procession of the sages. That office did not exist even in the imagination, till he came to fill it, and none can succeed to his place when he has passed away.[3]

Logical thought and imagination are two gifts that any good research scientist must possess, imagination, despite popular belief, being at least as necessary for the scientist as for the artist. Maxwell had them, but probably not in any greater measure than several of his contemporaries. What he did have in addition were flair and tenacity. He would mull over problems for long periods, disgorge his conclusions, but then, far from having finished with the subject, he would continue to mull over the same problems for years more, trying different angles of attack, to add another layer to our knowledge.

Flair is an indefinable but unmistakable quality; it shines through Maxwell's work. Peter Medawar has described science as 'the art of the soluble'.[4] At any given period of history there must always be major scientific questions that are insoluble: experimental techniques have not yet been developed that can provide enough background information and detail properly to begin to understand the problem, let alone solve it. One of the hallmarks of genius is to pick, intuitively, upon those major problems which are just within the bounds of the soluble at that epoch.

There is an element here of what Arthur Koestler prefers to call sleepwalking—of the kind performed by Harold Lloyd: of deciding on incomplete and sometimes actually misleading experimental data, what results are significant and what are, ultimately, irrelevant; of knowing instinctively just how far any theoretical model of phenomena can be pushed before it buckles at the seams.

A good example is Maxwell's problem with the specific heats of gases. The amount of heat required to warm a fixed quantity of gas through one degree centigrade is different if the gas is held at constant pressure or at constant volume. Remarkably, the ratio of those two distinct quantities gives a great deal of information about the internal structure of the gas molecules. Maxwell was able to predict one value of the ratio for monatomic gas molecules, and another value for polyatomic molecules. The experimental results fell neatly between the two: clearly an impossibility. So Maxwell had a theory, his statistical mechanical model of gases, which for the most part was wonderfully successful, but which broke down at this one small point. Trapped in such a dilemma, a scientist can say either that the experiment is wrong, or attempt to fudge his calculations, or he can admit that there is something deep going on that he does not yet understand. The temptation to take the first two options is understandably great. Here, however, Maxwell had the courage to stand by both the experiments and his calculations: he admitted that there was something wrong with the whole matter. The paradox of the specific heats was not, in fact, explained for a further forty years, and then required the revolutionary new theory of quantum mechanics for its resolution.

Being correct always seems easy in hindsight, and Maxwell's prescient quality stands out only in comparison with other scientists of the day. Maxwell exhibited a catlike awareness of which scientific debates to avoid. William Thomson, Lord Kelvin, was a contemporary of Maxwell, and a very great scientist, but much of the work that impressed his peers must now be discarded because it is based on false premises. One of Thomson's major areas of research was the age of the earth. Geologists and evolutionists required the earth to be old, to allow for the slow processes of geological change and evolutionary development. Kelvin leapt into this debate by saying that such long periods were impossible: no known source of energy could keep the sun hot for the boundless aeons envisaged by those scientists.

Thomson calculated that if the sun was powered by gravitational contraction it was probably 10 million years old, and at the outside 100 million years. True enough as far as it went, but Kelvin lived just long enough to have to eat his words, when the process of radioactivity was discovered. At a stroke a vast new source of energy was revealed, and ages of thousands of millions of years for the earth and the sun became possible. Perhaps that is the flaw that separates Kelvin from Maxwell. There is an inspired caution in Maxwell, that has resulted in very little of his work being superceded; very little has dated, and many of Maxwell's papers can still be read today, and profitably, as almost standard texts on their subjects.

For the rest of this book, we will explore, as far as it is possible, the various facets of Maxwell's life and work. We begin by looking at the historical background of science in Scotland, that threw up so many gifted researchers at around this time, and the particular traditions of Maxwell's family. Then we move on to Maxwell's schooldays, and his three years at the University of Edinburgh, both of which played major parts in forming his character. His formal education was completed at Cambridge, as an undergraduate and fellow at Trinity College. After five years, the mature scientist Maxwell left Cambridge to become professor of Natural Philosophy at Marischal College Aberdeen then on to King's College London before retiring to his family estate in Galloway. Aberdeen, London and Glenlair saw the bulk of his major research over a fourteen year period, but he did not tackle things sequentially; he would write a paper on electomagnetism, then do a number of other things, and perhaps return to electromagnetism five years later, to pick up where he had left off, with new ideas. Following the chapters on his life over this period, therefore, there will be separate chapters on his work, where the main themes of his reasearch will be followed through, logically, even though out of chronological sequence. Maxwell was tempted out of retirement, by appeals to his sense of duty, to become the first professor of experimental physics at Cambridge, and it was in Cambridge that he died in 1879. The final chapter is a discussion of Maxwell's impact on, and legacy to, science.

CHAPTER 2

If an undefinable, unquantifiable 'flair' is the hallmark of *genius*, it is nonetheless true that a lot of *very good* science does seem to spring up, almost overnight like mushrooms, in particular places and at particular times. Maxwell was the crowning glory of a 'flowering' of Scottish science, which had also produced Kelvin, Peter Guthrie Tait, Sir James Hall, James Forbes, Sir David Brewster, James Sutton, Charles Lyell, Joseph Black, John Playfair, John Robison, Joseph Lister, William Macquorn Rankine, John James Waterston, James Young Simpson, Balfour Stewart, James Watt. Thomas Young and Charles Darwin were by-blows of it. That background of talent helped Maxwell, sometimes directly—he was taught by Forbes and corresponded with Thomson—but also by creating a suitable scientific ambiance: a volcano towering out of a high plateau has to make less of an effort to get its head in the clouds than one rooted to the ocean floor.

The reason why the capital of Scotland, the Edinburgh of the Enlightenment, deserved the name 'Athens of the North' is much more susceptible to analysis than the individual genius of one man. Fig.1 shows that almost all the great names of nineteenth century British science belong to the North of England, Scotland and the North of Ireland. The names of southerners were not conveniently forgotten when drawing up the map: Faraday, Davy and Rayleigh were born in the South. It is difficult to think of many others. So perhaps the question of why Scottish science was so good should be inverted: why was southern English science so bad? After all, London and Oxbridge in the eighteenth century had housed the extraordinary flowering of talent of Newton, Hooke, Boyle, Wren and their friends.

The root of the trouble goes back farther than that. Oxford (and subsequently Cambridge) University was a religious foundation deliberately located outside the metropolis. In medieval times the church was a democratic institution, rewarding talent rather than birth (for obvious enough reasons). Since it carried out its business in Latin, which was unintelligible to the masses and therefore unlikely to lead to acts of violence in Grosvenor Square[1] its bright young men were allowed great freedom of thought. The result was that fourteenth century Oxford also saw one of those sudden flowerings of science; Heytesbury, Swineshead, Dumbleton and Bradwardine (who was also Archbishop of Canterbury) pushed medieval mechanics as far as was possible in those pre-mathematical days. They discovered the distance law for constant acceleration ($s = \frac{1}{2}(u + v)t$) or uniformly diform motion as they called it. This rule became known as the Merton Mean Speed Theorem in the medieval world because all four men were attached to Merton College.

Newton, too, was from a poor background, entering Cambridge on a scholarship, and finding his niche in academic religion. In some ways he was the last gasp of this medieval democratic tradition, for while there were always some scholarships to Oxbridge, after the Reformation, those universities became much more a country club, where the idle sons of the wealthy amused themselves and made useful social contacts. Instead of being open to all talent, the universities had been denationalized, and became the preserve of the privileged. But the fellows retained some monastic traditions: they had to take religious orders and remain unmarried, a much less attractive way of life than it had been been before the Reformation.

Being a fellow of a college was the only way one could hope to be paid, while doing science. Most scientists were amateurs, men with enough money or dedication to afford to have leisure, and, crucially, enough education to inspire them with interest in the natural world. England's empire was growing, demanding administrators to run it and soldiers to defend it. Administration promised wealth, and the army glory, and the English educational system, or at least its public schools, trained boys for those careers—spartan conditions, regimentation and fagging were deemed a suitable background for the army, Latin and Greek for administration. Science would have smacked of the practical, which would have smacked of trade—and that was a dirty word.

It is surprising then, that in the early nineteenth century, mathematics should form the foundation of the University of Cambridge education. The 'tripos' was an extraordinary phenomenon. It was for many years the one university exam: to get a degree one had to take the tripos, one had to study mathematics.

It arose partly as a homage to Newton, and partly because maths is a subject in which it is possible to set objective examinations. It is therefore typically perverse that this, the first written mass exam outside of Imperial China should be called the tripos, after the three-legged stool on which candidates used to sit when undergoing the previous examination system, trial by oral dissertation. To get a first in the tripos, to become a wrangler, was to get a meal ticket for life: the tripos syllabus was regarded as an excellent training of the mind for becoming a judge or a bishop. It was not regarded as a degree in mathematics. It was not even a particularly good training in mathematics. It emphasised speed and bookwork, the mechanical regurgitation of facts as rapidly as possible. Also, the dead hand of Newton lay on Cambridge mathematics. He had quarrelled furiously with Leibniz about which of them had been first to discover the calculus. Cambridge had backed Newton and saddled itself with his approach to the subject, fluxions, and Cambridge mathematicians were thus not trained in, and therefore incapable of using, Leibniz' more flexible approach, one of the main reasons why continental mathematicians in the late eighteenth century made major discoveries and Cambridge mathematics had fossilized.

While the law and the army were the preserves of the upper classes, academia the preserve of the church, and the south of England was a scientific desert, in the north of England science prospered. Many of its scientists came from non-conformist backgrounds. Outside the public school network, denied entry to Oxbridge, where the 39 articles of the Church of England had to be signed before entry was allowed, the non-conformists had to succeed by other means. They turned to trade and in the north of England, where both coal and iron ore were to be found, they turned also to technology. Many made their fortunes in steel, pottery or other manufacturing industries. Their children growing up in a technological environment did not learn to despise industry as did their public school peers to the south, and so the Joules, the Daltons, the Youngs and the Priestleys lived on their family fortunes, or were scientists in any spare time they could scrape from being doctors or teachers.

The religious connection is so strong in England that it has been suggested that non-conformism encouraged scientific endeavour in some more direct way. But this is difficult to maintain; Wales produced almost no science, while men of thoroughly orthodox belief in France (Catholic) and Scotland (Presbyterian) produced good science. The indirect effect of home and educational background is more likely to have been responsible.

Scotland was, like the north of England, a long way from the Imperial power base, and there too, more people had to work for a living, but the educational structure did not militate against science. Four of its five universities had been established within large cities: Glasgow, Edinburgh and two in Aberdeen, and so after the Reformation town played a large part in the future of gown. While theologicial studies continued, the more practical and vocational subject of medicine was introduced to the university syllabus. In Edinburgh it was pressure from the town council that helped to start the medical school: the councillors were worried about the drain of money and talent as young Scots went to Leyden to study, and thought to reverse both trends. Keeping abreast of advances in medicine also meant keeping abreast of advances in science generally. Joseph Black was a doctor who had a thriving practice in Edinburgh, but is today much better known for his fundamental work in chemistry, where he isolated and analysed carbon dioxide, and in physics, where he discovered latent heat: heat energy has to be put into ice to make it melt to water, or into water to convert it to steam, even though there is no change of temperature in either case. One suspects that the distillers of illicit whisky in the Highlands were well aware of that phenomenon—but by the nature of things, they were denied access to the scientific journals of the day.

Instead of the English emphasis on Latin and Greek, education in Scotland aimed to produce a 'lad o' parts', a good all-rounder with enough mental agility and self confidence to make his fortune in life. The only compulsory element in the university curriculum was philosophy, and in the hands of the dominant Common-Sense school, which tried to bring the certainty of science to philosophy, it became very much a philosophy of science. In Scotland, therefore, society's attitude towards science was much more favourable than in England.

The British Empire was engineered and doctored by Scots. James Watt was originally an instrument maker at the University of Glasgow, and a friend of Joseph Black. He appreciated the practical significance of Black's discovery of latent heat, and applied it to the very inefficient Newcomen steam engine then in use. By introducing the separate condenser, so that the steam did not sit in the piston taking forever to recondense, because of its vast store of latent heat, he began the industrial revolution. Later he emigrated to Birmingham, to the Soho foundry of the non-conformist engineer Matthew Boulton, where the skills to make large steam engines were developed, and Watt also became a leading figure in the Birmingham Lunar Society, a society of scientific amateurs, self-made men from the Midlands, who met once a month to discuss scientific matters. The name comes from the fact that the meetings were held at the full moon, so they could ride home after the meeting with a maximum chance of survival on the rutted farm tracks that in those days passed for roads.

The tolerant Scottish attitude towards science is well exemplified in Maxwell's family. The founder of the Clerk family fortune made his money in trade in Paris in the mid seventeenth century, and returned to buy the barony of Penicuik. This outward-looking example was followed by his grandsons, who both went to Leyden to study under Herman Boerhaave, the remarkable man who revolutionized medical education by showing his students real live patients, rather than books, and inspired the founding and the methods of the medical school at the University of Edinburgh.

The older of the two, John (later Sir John) Clerk, proceeded to Italy, to complete his education in the arts and music, and to hear an opera he had written performed in Rome. He had lost none of the family business acumen, however; on his return to Scotland he became Baron of the Exchequer and was a commissioner of the Union. The other grandson, William, married Agnes Maxwell, the heir to the vast Middlebie estate in Galloway, but he had to change his name to Clerk Maxwell to get his hands on it.

All three of Sir John's sons maintained the family's scientific tradition. The two eldest, James and George studied at Leyden, and George and the youngest brother, John, were both interested in geology. They were friends of Sir James Hutton, whose book *The Theory of the Earth* (1788) is generally regarded as the beginning of geology as a science. In it Hutton rejected the biblical chronology of the earth, and the flood as the sculptor of the earth's present surface. This was known as the catastrophe theory. Instead he pictured the 'boundless pre-existence of time' as Black wrote,[2] an age-old earth where slow geological processes ground on, rain water wearing down what volcanic action piled up. One historian of geology, perhaps a little biased, placed Hutton alongside Newton as being 'one of those rare scholars who open an entirely new field to the human mind. What one did for Space the other accomplished for time.'[3] Hutton's followers became known as the Uniformitarian school, and they were to incur the wrath of Lord Kelvin through their demand for vast swatches of time in which geology could take place.

Hutton was not merely the greatest theoretical geologist of all time, he had trained as a doctor, and then practised farming for fourteen years before retiring to a life of science in Edinburgh. And there he was an entrepreneur too, making money out of a partnership to extract sal ammoniac from the city's soot, an early form of recycling.

One of Hutton's major pieces of evidence for uniformitarianism was that granite (i.e. volcanic) mountains are pushed up through the surrounding, and therefore older, rock. He made a number of famous geological forays around Scotland collecting evidence for his theory, in particular to Glen Tilt and to Arran, and the Clerk brothers accompanied him on these journeys. Indeed, John Clerk was a splendid draughtsman and water-colourist, and produced some paintings of those structures, granite intrusions, intended to illustrate the third volume of Hutton's book. Unfortunately the book was never published, and the drawings were lost for almost two hundred years, until discovered in 1968 in the Clerk family's papers at Penicuik.[4]

The Clerks' enthusiasm for geology made sound business sense. The family had coal mining interests in the Edinburgh area. John's interest had been sparked off by these when he was a boy. He later wrote, 'from the very reason of it's being conceald (it) might engage a young mind to be very curious and even sollicitous . . . to understand the nature of it'.[5] Even more curious is the way this almost exactly parallels the boyhood experience of James Clerk Maxwell, and his father, John Clerk Maxwell. His forebear John Clerk, of Eldin, was a splendid character. He was passionately interested in naval tactics, and wrote a book on the subject which he claimed enabled Admiral Rodney to win the naval battle off Dominique in 1782. He used to practise his tactics with flotillas of model boats in the fish ponds at Penicuik.

George Clerk's business instincts were eventually to get the better of him. He indulged in some speculative mining operations in Lead Hills and elsewhere, which went disastrously wrong and lost him a fortune. His father, Sir John Clerk, had married him off to his cousin, Dorothy Clerk Maxwell, who had been orphaned at the age of seven, and had become Sir John's ward. As she was the heir to the Middlebie estate, it had the welcome effect of keeping the estate in the family. It was still encumbered with its original Maxwell debts, however, and Sir John, in an operation that can be described only as crafty, went to Parliament, sold off the entailed estate to pay off those Maxwell debts, bought it himself, and settled in on George, who became George Clerk Maxwell. With his mining speculations collapsing in his face, however, poor George was forced to sell off most of the estate, and the only part left was Glenlair, an estate of 1500 acres in Galloway. George, having lost Middlebie, then gained Penicuik, when his elder brother died without children, and it was his second grandson, John Clerk Maxwell (1790-1856) who was the father of the physicist, James. John inherited Glenlair and also the family's scientific bent.

He went to school and to university in Edinburgh, and then became an advocate (the Scottish equivalent of a barrister). He does not seem to have been particularly interested in the law, but the life was very congenial, and the vagaries of family inheritance had left him comfortably off. In between cases, the Parliament House (the law court) was an excellent meeting place where intellectually and philosophically minded young men with nothing better to do could meet and talk. Maxwell's main interest lay in technology. As a boy he had been inspired, as had his great uncle John Clerk of Eldin, by watching machinery, though in his case it was a paper-mill near Penicuik. As a young man, he tried to devise a sort of bellows that would produce a smooth continuous blast of air (as is absolutely crucial to the operation of all blast furnaces today), and in 1831 he wrote a paper 'Outlines of a Plan for Combining Machinery with the Manual Printing Press'. He seems not to have proceeded further with this scheme for an automatically-fed press than publishing this paper—perhaps new technology was as unpopular with printers then as it is today.[6]

His partner in revels was John Cay. Together they discussed his wild schemes, but their favourite occupation was attending meetings of the Royal Society of Edinburgh, in those days much more all-embracing than now. In addition to topics of pure science, schemes for lightning protection, geographical exploration and education would be discussed at the regular Monday evening meetings. The listeners would sit on cushioned benches around a large central table on which the lecturer of the evening would display his wares. Science was on a much more human scale than today: instead of massively expensive immovable equipment, or the statistical analysis of a vast number of complicated experiments, much more could be demonstrated directly, to the enjoyment and edification of the spectators. Joseph Lister, the surgeon who introduced antiseptics to the operating theatre, on one celebrated occasion was to elucidate Pasteur's theory of germs by bringing in a glass of milk which he had sealed hermetically months before in preparation for this great event, and drinking it in front of his astonished audience. Maxwell thus whiled away his time pleasantly enough until he was thirty six (1826) when he married Frances Cay, the sister of his mechanical friend. She was then thirty four. Their only child—a daughter Elizabeth had died earlier—James Clerk Maxwell was born on 13 June 1831 at 14 India St. in Edinburgh.

With a father like John Clerk Maxwell, there was little danger that James would be educated away from science. The relationship that developed between the technologically-minded father and the scientifically-minded son is a fascinating one. In James' formative years, the father had the patience, leisure, knowledge and enthusiasm to answer the boy's eager questions about the inner workings of the universe, or the bell-pulls in their house. At school he encouraged and pushed James on in his work. Then as the adolescent James spread his wings, their roles reversed, the son explaining his latest ideas to his father.

There is a touching letter sent by John to James, who was then taking a short holiday in Birmingham while working up to the tripos. He was staying with his college friend Johnson Gedge, whose father, the Rev. Sydney Gedge, was the Second Master at King Edward's School:

Edinburgh, 13th March 1853.
Ask Gedge to get you instructions to Brummagem workshops. View, if you

can, armourers, gunmaking and gunproving—swordmaking and proving—Papier-mâché and japanning—silver-plating by cementation and rolling—ditto, electro-type, Elkington's Works—Brazier's works, by founding and by striking up in dies—turning—spinning teapot bodies in white metal, etc.—making buttons of sorts, steel pens, needles, pins, and any sorts of small articles which are curiously done by subdivision of labour and by ingenious tools—glass of sorts is among the works of the place, and all kinds of foundry work—engine-making—tools and instruments (optical and philosophical) both coarse and fine. If you have had enough of the town lots of Birmingham, you could vary the recreation by viewing Kenilworth, Warwick, Leamington, Stratford-on-Avon, or suchlike.[7]

The touching thing about this letter is the revelation of the inevitable gulf that had opened between father and son: the father's feverish list, trying to prove he still had an advisory role to play in his son's life. But James' main interests had long left technology for fundamental science, and one suspects that Stratford-on-Avon would have been much closer to the top of his personal list, since he deeply loved and had an intimate knowledge of Shakespeare's work. But ever the dutiful son, James began with the glass works.

CHAPTER 3

Soon after their marriage, John and Frances Maxwell decided to leave Edinburgh to make their permanent home at the family seat of Glenlair. In those pre-railway days, Glenlair must have seemed to be the back of beyond. It was a bone-shaking two-day journey by carriage to get to Edinburgh and civilisation, but when Mrs. Maxwell became pregnant the couple had to make the journey so that medical attention could be close at hand, if it proved necessary. The decision to move away from all the comforts of life may seem a strange one for the rather 'indolent' (Campbell's description) John Clerk Maxwell. Partly, no doubt, it was his wife's influence: 'she was of a strong and resolute nature,—as prompt as he her husband was cautious and considerate,—more peremptory, but less easily perturbed', but partly it was also an unexpected streak of the pioneering spirit in him.

Today Glenlair stands in beautiful countryside. Galloway is a land of gently rolling hills, small fields, hedges, woods and water—purling streams and island-strewn lakes. It has the man-made softness of the English countryside rather than the ruggedness associated with Scotland (whose beauty is in general sheep-blighted more than man-made), as if a small chunk of the Lake District had become detached and floated to the opposite side of the Solway Firth.

Indeed, Galloway looks that way because it *is* man-made. For centuries the Border country between England and Scotland was the canvas on which a horrific tapestry of robbery, murder, kidnapping, rape and feud was woven. The clans on both sides of this no-man's land set standards of viciousness and deceit which the Mafia would be proud to emulate. While England and Scotland existed in a state of hostility, if not actual war, it suited both countries to turn a blind eye to the activities of those clans: they were a constant drain and irritant on the enemy, and they were an invaluable guerilla force striking at the lines of communication behind an invading army. If the Borderers feuded with their compatriots as much as with a notional enemy, well, they lived outside the law, and no-one cared. Under the circumstances, it is hardly surprising that the Borderers evolved to be tough, crafty, ruthless and clannish.

The most vicious vendetta of them all was between the Maxwells and the Johnstones: neighbouring Scottish clans at the extreme Western edge of the area of conflict, which enabled them to devote all their courtesies to each other. The net result of all their rape and pillage was that it became virtually impossible for an honest farmer to earn a living, and the land was reduced to a gorse-covered desert. In 1528, Lord Dacre wrote to Cardinal Wolsey that the Maxwell-Johnstone feud had turned the 'Debateable Land' (the Border areas) into a waste. The feud culminated in the Battle of Dryfe Sands in 1593, when the Johnstones, threatened with annihilation by a numerically superior Maxwell army, fought and won a desperate last-ditch battle. John, the eighth Lord Maxwell, was killed. As a coda to this story, in 1608 Lord Maxwell and James Johnstone met to effect a reconciliation. Precautions had been taken to ensure a peaceful meeting, and Maxwell promptly shot Johnstone twice—in the back. Maxwell had to flee the country, and one of the most beautiful of all Scottish poems, 'Lord Maxwell's Goodnight', is a heroic and passionate account of his farewell to his former life. It makes no mention of why he had to disappear so rapidly.

By that time the crowns of England and Scotland had been united, and an uneasy peace had descended on the borders, although casual rustling continued in a minor key to within living memory of the time of James Clerk Maxwell's birth. The land was slow to recover from its legacy of strife: religion added to the soil's burden of blood. Presbyterians in the middle seventeenth century and Covenanters in the late seventeenth century were murdered impersonally.

The Maxwells had grabbed the Middlebie estate in 1666, when its owner, John Neilson, disgusted with being persecuted as a Presbyterian—fines, his cattle impounded and sold, troops quartered on his property requisitioning their provisions from him—joined some like-minded gentlemen of the district and rose in rebellion. It was crushed, he was given the fashionable torture of the day, the 'boot', and then hanged. Once the Maxwells had occupied the property, they made certain of keeping it, and wrote into the entail of the estate that whoever inherited the property had to take on the Maxwell surname. Until the nineteenth century he had also to take on the legacy of a blighted and impoverished landscape. Then some farmers started to clear, plant, sow and reafforest the land. The most eloquent amongst them was Thomas Carlyle, who wrote to Goethe in 1828:

> Our residence is not in the town itself (Dumfries), but fifteen miles to the north-west, among the granite hills and the black morasses, which stretch westward through Galloway, almost to the Irish Sea. In this wilderness of heath and rock our estate stands forth a green oasis, a tract of ploughed partly enclosed and planted ground, where our corn ripens and trees afford a shade, although surrounded by sea-mews and rough-woolled sheep. Here, with no small effort, we have built and furnished a neat, substantial dwelling.[1]

His wife, Jane Welsh Carlyle was not so keen on the back-to-nature existence, describing the area as 'a waste prospect of heather and black peat moss, with not a shrub to prune', and six years of Gallovidian solitude did not improve the view; as Carlyle's biographer wrote:

> So still the moors were, that she could hear the sheep nibbling the grass a quarter of a mile off.[2]

It eventually drove the Carlyles down to Chelsea, where Thomas was able to romance in comfort about the peasant existence.

John Clerk Maxwell was one of the pioneers who managed to stick it out, and by dint of strenuous effort, to restore much of the land to its natural wooded beauty and fertility. The Maxwell family thereafter could never bring themselves to indulge in Carlyle worship, then a popular intellectual game. Later on, James was obliged to take a worshipper, who also happened to be a houseguest, on an excursion to Carlyle's 'shrine', his house at Craigenputtock. Maxwell was fond of telling what happened next:

> The enthusiast, in his rapture, harangued an old peasant, who was hoeing "neeps", on the glorious doings of the former tenant of the farm-house. The man listened, stooping over his work till the rhapsody was over, then looked up for a moment saying, "It is aye gude that mends," and resumed his labour.[3]

John Clerk Maxwell found much to mend at Glenlair. When he and his wife arrived, there was no house to live in, and the fields had to be cleared of stones. All his mechanical ingenuity found a natural expression; he surveyed the land, designed the buildings, and supervised the masons who erected the house. Family legend recounts that when he returned from Edinburgh with his wife and infant son, he planted the seed of a copper beech tree in his garden— and a splendid specimen is still flourishing there today. It would be especially fitting if the story was true, since the copper beech is not only a beautiful tree, but it is also one which botanists then had discovered only recently, growing as a sport in Kew Gardens—what better symbol for the reclamation of Galloway?

John Clerk Maxwell's pioneering spirit extended to more than agricultural science and house building, it even covered the sartorial arts. He disapproved of the current fashions, and had sensible square-toed shoes (offering ample room for the toes) made for James and himself by the local shoemaker, from leather he had personally chosen. He acted as tailor, cutting out his and James' shirts to a pattern of his own design. This may seem a minor eccentricity, but it was one which was to have a profound impact on James' life: his quaint clothing strongly affected his school career. On the other hand, the practical experience in building and surveying gained while accompanying his father on his rounds, eventually was to make the Cavendish Laboratory in Cambridge a model of laboratory design, a building which the architect was given no chance to blight, but whose functional design reflected Maxwell's own experience both as scientist and builder.

The Urr valley in which Glenlair stands, was known to its residents as the Happy Valley, because a number of sociable families had made their homes there and they entertained each other to picnics, afternoons of badminton and archery in the summer, skating and curling in the winter. But for young James, there was no-one of his own age and class. For much of the time he played with the farm labourers' children, and the rustic Gallovidian accent he acquired was to be a permanent legacy. It is a measure of his independence of character that he made no attempt to lose it through his school and college days. (He once humourously remarked in a lecture on the latest marvel—the telephone—that he should have gone to Alexander Melville Bell, father of Alexander Graham Bell, a teacher of elocution in Edinburgh, who 'though a Scotsman . . . taught himself to speak English in six months'.)[4]

He also never lost a paternalistic concern for the welfare of the working man. His letters home from school always inquired about his friends, and he prescribed an improving course of reading for them. At Cambridge, he became involved in the Christian Socialist Movement, and for some years after gave up an evening a week to deliver lectures on scientific topics to working men who wished to better themselves. He later took his duties as 'laird' of Glenlair with the utmost seriousness, once offering to build and maintain the village school at his own expense, and conscientiously visiting his sick labourers to pray with them. In those more religious days, that was not so patronising as it might appear today, and seems to have been genuinely welcomed by the recipients. There was, nonetheless, a clear social division in his attitudes. He referred to the local people as the 'vassals', and on one occasion invited a friend to stay saying:

> I can promise you . . . a reasonable stock of natives of great diversity, and very unlike any natives I know elsewhere.[5]

One of his brightest students, Donald MacAllister, captured Maxwell exactly when he described him as a 'thorough old Scotch laird',[6] he had the attitudes as well as the ingrained sense of obligation that the phrase implies.

Apart from his parents, the person who seems to have been closest to James in his childhood was his cousin Jemima Wedderburn, who was eight years older than him. She lived in Glasgow with her widowed mother but spent a lot of her holidays at Glenlair. She was also a talented artist, and her water colours, still clear and fresh, preserved in an album by the family, tell more about life at Glenlair than anything else (Plates I-XX). In them, 'The Boy' as James was referred to, is seen leading his gang—of Jemima and the labourers' sons—on climbing expeditions on the rocky banks of the Urr, out playing with Red Indian headdresses, standing silently observing Happy Valley occasions. One picture is of basket-weaving by rushlight, the whole family, and their servants, gathered round a long table finding useful activity for a winter's evening. Another picture shows the harvest-home celebrations, everyone gathered in a large barn, and the six year old James, long past his bed-time, staring up intently at a worried-looking fiddler. Yet another shows James wearing the remarkable clothes his father had designed, teaching his dog, companion as well as pet, to perform tricks.

The abiding impression conveyed by the paintings is of the concentration and independence of the little boy in his Bob Dylan cap, whether he is participating in weaving a basket, or watching from the outside the strange antics of adults. Being an only child, he must very often have been lonely living in the country.

The relationship between father and son was extremely close. All children are naturally inquisitive and ask questions endlessly. James was no exception. His standard question was "What's the go o' that?" If he was unsatisfied with the precision of the result, this was to be immediately followed up with "But what's the *particular* go of it?" Infuriatingly and inevitably, his chain of questions would soon terminate in the ontologically unanswerable: once he held two pebbles in his hand, a piece of red sandstone, and another of blue whinstone. On being told that the whinstone was blue, he asked, "But how d'ye know it's blue?"[7]

Amusing though it might be to try to trace his lifelong interest in optics to this moment, there is nothing remarkable about this: all children ask questions. What was remarkable was his father's response. The normal reaction to the inquisitive child, if he is lucky, is that of one of James' aunts who, exhausted by a prolonged verbal battering, remarked how humiliating it was to be asked so many questions one could not answer by 'a child like that'. If the child is unlucky he gets a quick blow. But James was singularly fortunate in his choice of father. Here is his fond mother writing of James at two and a half years old:

> He has great work with doors, locks, keys, etc and "Show me how it doos" is never out of his mouth. He also investigates the hidden course of streams and bell-wires, the way the water gets from the pond through the wall and a pend or small bridge and down a drain into Water Orr, then past the smiddy and down to the sea, where Maggy's ships sail. As to the bells, they will not rust; he stands sentry in the kitchen, and Mag runs thro' the house ringing them all by turns, or he rings, and sends Bessy to see and shout to let him know, and he drags papa all over to show him the holes where the wires go through.[8]

The image of the tiny boy dragging his fifteen-and-a-half stone father round the house is a delightful one. Yet his father did allow himself to be so bullied. He had the technical knowledge, the leisure and the interest to answer all but the most rigorous of James' questions, and even to direct the boy's interests. James' nanny, Maggy, recalled one incident at about that time when she had given James a new tin plate to play with:

> It was a bright sunny day; he held it to the sun and the reflection went round and round the room. He said, 'Do look Maggy, and go for papa and mama.' I told them both to come, and as they went in James sent the reflection across their faces. It was delightful to see his papa; he was delighted. He asked him, 'What is this you are about, my boy?' He said, 'It is the sun, papa; I got it in with the tin plate.' His papa told him when he was a little older he would let him see the moon and the stars, and so he did."[8]

A celestial globe was found preserved at Glenlair with the constellations cut out by hand to make a stellar jig-saw puzzle: his father's brilliant piece of educational psychology in following up his son's first flush of astronomical enthusiasm. The fact that James was to preserve it intact is equally typical and significant.[9]

It was his mother, though, who attended to James' early education. She was a religious woman and taught him long passages from the Bible. That may well have helped to train his phenomenal memory: his ability to remember appropriate references was a great help to him in science, and his facility in dredging up appropriate quotes for all circumstances from the Bible, Shakespeare and other sources, not only astonished his contemporaries, but also effectively screened the cogent points he was actually trying to make. By the age of eight, his mother was able to boast, James could recite all 176 verses of Psalm 119, the longest in the Bible. Four of the lines of the Psalm have a curious resonance for Maxwell's career:

Open thou mine eyes,
That I may behold wondrous things out of thy law.
I am a stranger in the earth:
Hide not thy commandments from me.

But Mrs. Maxwell's pleasure in her son's achievement was short-lived. She died in 1839, aged forty-eight; the blow to James and his father was shattering. Its immediate effect was to drive them even closer together. Jemima's pictures now show James dressed in a miniature version of his father's dress, blue tartan trousers, solemnly sitting on his pony following his father's phaeton, to view recent 'improvements', to watch ploughing matches, to survey land for building, and stand by while his father shot rabbits.

His father delayed sending James to school for as long as he dared: he enjoyed his son's company and so brought in a private tutor, a local sixteen-year old lad, with no experience, who was persuaded to postpone going to university in order to give James lessons. This proved a minor tragedy. The tutor was the product of a harsh educational system, and he relied on the tawse (a stout leather belt), the pulling of ears till they bled, and the battering of the head with a ruler to instil educational values (particularly unquestioning obedience). James was already an independent-minded boy and simply refused to learn anything at all under that regime, and refused to answer any questions. Campbell attributes Maxwell's later 'hesitation of manner and obliquity of reply'[10] to the treatment he received. Maxwell was to adopt a perverse way of actually answering questions while appearing not to: his answers were a sort of verbal crossword, that seemed to ramble on irrelevantly until the listener either gave up, or spotted the key and then they made excellent sense.

Maxwell gained one glorious moment of revenge on his tutor, which Jemima captured forever. Goaded beyond endurance one day, he climbed into a washing tub and paddled out into the middle of the duck pond and refused to return. Instead of running away and being simply ignored, he made his defiance fully public, sitting in the pond just out of reach from the bank, to the obvious enjoyment of the 'vassals', Jemima and his father. James made no direct protest, however, and it was left for his aunt, Miss Cay, to divine the true state of affairs—the tutor had pronounced that James was a slow learner. The tutor went, and James had to be sent off to school.

One of the joys he had acquired in his prolonged spell of rustication, was a profound love of Nature, gained mostly through his solitary pursuits around his beloved Urr. Unusually, for a country-bred boy, he never shot or fished. He preferred instead to watch and admire the wonders of Nature, to blow irridescent soap bubbles and watch them shimmer, to study the abstract sculptures the Urr cut in its rocky bed, or the foam whorls and eddies it piled up when in spate. His tub became a vessel of exploration, to watch ducks and moorhens in their nests, and to observe his favourite animal of all—the frog. He used to pick up frogs—'Clean dirt' was his generic description of anything that crawled in which he was interested—stroke them, hear them croak, pop them into his mouth and let them jump out again to terrify his parents.[11] His frog imitations were to become famous. Maxwell had an uncanny way with all animals. He rode a horse that everyone else had agreed was untameable, and had a succession of dogs, all of whom he taught to perform circus tricks. His predilection for the frog was not a mere perversion, however; the frog's transformation from egg to tadpole, and from tadpole to frog is one of Nature's miracles, not least because it occurs all around, in lowly ditches and puddles, where it is generally completely ignored.

Through his observations of the frog and other neighbourhood wildlife, Maxwell took a major step towards becoming a true scientist, observing without prejudice the real facts of Nature, and never losing the sense of wonderment at her miracles. He always felt that this pursuit, if it was to mean anything, had to be self-taught and pursued alone; a philosophy he was later to follow when guiding the research of others at Cambridge.

Campbell recalled the deep significance Maxwell attached to Burns' lines:

The Muse, nae poet ever fand her,
Till by himsel he learn'd to wander;
Adown some trottin' burn's meander,
    An' no think lang. [12]

They show Maxwell's feeling for the Urr, as well as for the deep wellsprings of the imagination.

After the debacle of the tutor, playing by the stream, tadpoles and tubbing had to become summer pursuits, for most of his next years were spent in Edinburgh, staying with his aunts Miss Cay and Mrs. Wedderburn and being educated at Edinburgh Academy.

CHAPTER 4

Edinburgh Academy had been founded in 1824 by Sir Walter Scott, Lord Cockburn and a group of like-minded Edinburgh New Town intellectuals. Its first Rector (Headmaster) was the Reverend John Williams, a Welshman educated at Oxford, and a classicist who had taught previously at Winchester and who had been tutor to one of Walter Scott's sons. The school's remit was to teach all the 'branches of study which are essential in the education of a young gentleman', and that meant concentrated doses of classics which would enable them to compete with the products of the English Public School system for places at Oxford and Cambridge, in the Civil Service and in the army.

How well Williams succeeded in his appointed task can be gauged from the career records of the boys in Maxwell's class. (Williams left the school in 1847, the same year as Maxwell.) Of the ninety-one boys in the year (not all of whom served their full sentence of seven years), three died young, and for twenty seven there is no record. Of the remaining sixty-one, twenty eight served in the armed forces, or went to India, or both; a further fourteen went abroad to other parts of the World.[1]

The Table of the school syllabus shows how this success was achieved. There is an emphasis on Latin and Greek and mathematics (as opposed to arithmetic) began with Euclidean geometry in the 5th class, when the boys would be, on average, fourteen. One must feel considerable doubt about the value of the arithmetic taught beforehand, judging by the curious value the Table gives for the total number of hours' teaching in the Second Class. The little physics that was taught began in the 6th class as an adjunct of Latin, and then mysteriously became a sub-branch of English in the 7th Class.

The school laid enormous emphasis on anglicising its pupils:

> The English Class is viewed by the Directors as of peculiar importance. It is conducted by a gentleman of English birth, well-versed in the English language and literature, and in the sciences of Grammar and Elocution, whose whole attention is devoted to training the younger classes in reading and spelling, and in the grammar, etymology, and structure of the language, and in leading on the higher classes to analyse the works of Milton, Shakespeare, and other classic authors, and to the practice of English Composition. The Pupils have thus the best opportunity of acquiring a correct pronunciation and intonation, habits of accurate composition, and a taste for the excellencies of English literature.[2]

In the face of pressure like that, the way in which Maxwell clung to his

Gallovidian accent becomes rather heroic.

For the first four years of a boy's school career, his 'class' teacher took him for all his classics lessons and that accounted for two-thirds of his school existence. After that the Rector applied the final polish in Latin and Greek to the boys who had survived. And it was a matter of survival, not merely physical—two of the class died while at school—but even more so, emotional. In a class of sixty-eight boys more jungle warfare than Latin is taught. The classmaster's main job must have been trying to maintain a modicum of control, with the tawse as his main ally.

Maxwell's classmaster was Mr. Carmichael, about whom little is known. But the class below was taught by Mr. Cumming, and the boys in the class formed a club, the Cumming Club, which continued to meet for years afterwards, and published a fascinating book, *The Chronicles of the Cumming Club*, about its members' school experiences, and their subsequent careers. Mr. Cumming was so good a teacher that many parents of boys who should by age have been in Mr. Carmichael's class, delayed sending their offspring to the school for a year, to allow them to get into the Cumming Class. Maxwell was one of the youngest boys in his year, younger than a number in the Cumming Class, and yet he joined the Carmichael Class in its second year. It would have been more sensible for him to have started his school career with all the other new boys in the Cumming Class, so, presumably, that must have already been over-subscribed by the time Maxwell was enrolled at the Academy.

Not that Mr. Carmichael was anything but a perfectly competent teacher. In addition to Maxwell, the class also produced Lewis Campbell, Maxwell's friend, eventual biographer and Professor of Greek at the University of St. Andrews, Coutts Trotter, traveller and explorer, and Charles Robertson, who was ordained after taking his degree at Cambridge where he had collaborated with Maxwell on his research into optics. The class also contained the unfortunately named H. Suetonius Officer, about whom the school records can only say cryptically 'went to Tasmania'.

The Cumming Class of the following year was even more remarkable. From it came P.G. Tait, a friend of Maxwell and Professor of Natural Philosophy at Edinburgh and Fleeming Jenkin later to be Professor of Engineering at Edinburgh,—he collaborated closely with William Thomson (Lord Kelvin) in the profitable venture of laying telegraphic cables, and also worked closely with Maxwell in standardising the measurements of electrical quantities—and his forceful criticism of the Theory of Evolution drove Darwin into Lamarckianism. The class also contained Edward Harland, who did not go to university when he left the school, but was apprenticed to an engineer—and eventually rose to become a founder of the Harland and Wolff shipyard in Belfast; and Patrick Heron Watson, who achieved fame as a surgeon in the Crimean War.

Many of those illustrious old boys of the Academy are associated with scientific subjects. Despite the importation of English ideas on the importance of classics, many of the teachers at the school were homegrown products of the Scottish educational system, none more so than James Gloag, the maths master. The *Chronicles* devoted a whole chapter to him, so

large did he loom in the minds of his former pupils. Partly it was
that in a school of hard-hitting teachers, he hit the hardest. P.G. Tait
attributed this to his mathematical ability:

> Gloag could get more work on the tawse than could any of the other masters.
> His secret was in great part a dynamical one.[3]

But there is no doubt that Gloag also loved his mathematics, and, like Hartopp
in Kipling's school stories, would in no way tolerate his subject being belittled
for the sake of a dead language like Greek.

The point is made in one of the anecdotes in the *Chronicles*, that Gloag
wrote up a geometrical proposition on the blackboard, one Saturday morning
when the Rector (nicknamed 'Punch') was sitting in on the lesson. Gloag asked
round the class to see if anyone could prove it. None could do it till he got to a
boy named Sellar, the best classicist in the school and consequently the
Rector's favourite. Sellar too found it difficult:

> "Noo, Sallar," says Gloag, with a tap on the board, "don't keep us waiting on
> ye all day."
> Still there was no response.
> "Why, Sellar, my boy," says the Archdeacon, disappointed, "don't you see it?
> Think a moment—it's quite easy, don't you know?—perfectly simple."
> Here is the moment of triumph, so skilfully approached by Gloag, who,
> bursting out like a thunderbolt, exclaims—
> "Naw, Mr. Rector, sir, it's *nott* easy—the thing is impōssible, sir; its grŏss
> nonsense, sir!"[4]

The reviewer of this book for the *Edinburgh Academy Chronicle* in 1922 threw
doubt on the strict veracity of this story (it certainly is true that the book's
author, Colonel Fergusson, seems to take malicious delight in anything
redounding to Punch's discredit). But the point is that a lot of former pupils
of the school *did* accept the story as at least retellable even if not absolutely
true: the psychology of the story must be true even if not all the facts are.
While Gloag was at the Academy maths was going to remain an important
element in the boys' education.

There is no doubt, however, that the science teaching in the school was extremely
weak. As well as the evidence of the syllabus, we have Campbell's personal account:

> To keep our education "Abreast of the requirements of the day", etc., it was
> thought desirable that we should have lessons in "Physical Science". So one of
> the classical masters gave them out of a text-book. The sixth and seventh classes
> were taught together, and the only thing I can remember about these hours is
> that Maxwell and P.G. Tait seemed to know much more about the subject than
> our teacher did.[5]

The *Chronicles* tells us how the Cumming Class, at least, gained their real
science education. The boys formed a 'Philosophical Society' which met regularly
on Saturday mornings in Fleeming Jenkin's house. Geology was their favourite
science, as the search for fossils and specimens were real expeditions, leading them
to exotic locations, such as a limestone quarry just south of Edinburgh:

> The mines ran for a long distance underground, and were pitch-dark and, to
> reach the point where the miners were at work, it was necessary to come

provided with candles. A procession of small boys with lighted candles, feeling their way along the dark galleries, must have been an interesting sight, no doubt. They were always welcomed by their friends, the miners, at the far end; and well taken care of. For when there was going to be a blast, or what they called a 'shot' fired, the workmen would see that each boy was carefully hidden away, crouching behind a stone, or placed out of harm's way in lee of a projecting rock. The candles were put out for the moment while these precautions were taken. Then a blinding glare of fire, and a terrific crash echoed and re-echoed, which seemed to shake the solid walls of the mine; and for a few seconds the fall of fragments of rock and stones in the darkness all around formed a most impressive scene.

. . . The pursuit of Science on foot, in distant country localities, is hungry work; and banquets of oatmeal cakes, cheese, and ginger-beer in certain village change-houses are bright spots in memory.[6]

In the winter, they indulged in their second favourite subject, chemistry, with the ever-present possibility of producing obnoxious smells. One 'philosopher' once suggested an experiment to determine the result of putting a pound of potassium in a pint of porter. He was dissuaded. If there were no experiments, one of their number was deputed to write a paper on some topic, which would be mercilessly torn to pieces by the others. A 'Philosophical Society' like that, comprised of keen, bright boys following up whatever chance and interest led them to, must be close to an ideal introduction to science. It made up in interest what it lacked in comprehensiveness, and the ground rules of science could be picked up at a later date.

As Fergusson sadly noted, however, such a society could flourish only in the early days of the Academy, before Saturdays too were taken away by the school to be devoted to those twin public school deities, Cricket and Football. In many ways Maxwell was lucky to go to the Academy when he did; it was already an established school, but still youthful enough for traditions not to have hardened into a rigid cast. The pressure on bright boys to concentrate on the classics was not yet overpowering. The beginning of the end can be seen in the note at the top of the school syllabus, where it says that changes will soon be introduced in the upper classes' courses: even more Latin and Greek to get them into Oxbridge.

If the general climate in the school was not unfavourable to Maxwell's development, the specific circumstances of his arrival could hardly have been worse. He was thrown to the lions in the second month of the second year into a class of sixty-seven boys, most of whom had already had a year of each other's company, to form cliques and alliances. He had been brought up in the country, with very little company of boys of his own age. He was somewhat shy and in a crowded classroom where forwardness and speed were the criteria for survival, replied to questions only indirectly. Tait later recalled that Maxwell 'was at first regarded as shy and rather dull'.[7]

Maxwell also had a strong regional accent, and he appeared on his first day wearing the 'sensible' but far from elegant clothes and shoes his father had designed. Square-toed shoes with clasps, a soft frill round the neck instead of a starched collar, a tweed smock instead of a cloth jacket; all eminently practical, but all different.

He returned home after his first day at school with his clothes in tatters. His slowness, hesitancy and indirectness of speech earned for him the nickname 'Dafty', and his classmates thought him a genuine rustic halfwit. The nickname did not die quickly; cramming his brain with irregular Latin verbs, seemingly without rhyme or reason, with no feel for Latin as a language being developed, had no appeal at all. His hesitancy and shyness merely deepened.

During breaks he did not much enjoy noisy gang activities; he was a country boy and used to solitary pursuits. His only outlet was to climb the few withered trees that remained in the school yard, with their roots clinging precariously to the slimy banks of the Waters of Leith, the stream that runs through Edinburgh.

> The 'Waters of Lethe,' the wits of the Academy delighted to call that not too savoury stream . . . It was usually little better than a succession of stagnant pools, choked with weeds and other matters.[8]

It was a sorry replacement for his beloved Urr. Fegusson also remembered him, in a cloud, watching butterflies in the sunshine, while the rest of the boys played 'knifey', a complicated version of splits which involved impaling a knife in the ground between or near one's opponent's feet in various tortuous ways.

Maxwell's love of Nature brought him further misfortunes from the older boys in the school, the 'tyrants' Tait called them:

> Among Maxwell's peculiarities, which led us in our ignorance to dub him "Dafty," was the habit he had acquired of squatting and jumping in exact imitation of a frog. When this became known to the tyrants, "Dafty" was constantly called upon to make sport for them, jumping frog-like over a handkerchief, under the stimulus of the lash, till his exhaustion was complete.[9]

The first two years at the Academy must have been agony for Maxwell. The earliest letter home which Campbell prints in the biography has a desperate final paragraph, appealing for titbits of news about what is happening in the 'real' world back at Glenlair:

> Does Margaret play on the trump still? and what are the great works? Does Bobby sail in the tub?—I am your obedient servant
>
> > James Clerk Maxwell [10]

In retrospect, Maxwell's trials at school hardened him, made him more independent and self-reliant than his domestic educational experience had done. Someone of weaker character would have been destroyed, but as a recipe for nurturing *genius*, Maxwell's early education is hard to beat, for there independence and self-confidence are two of the most vital ingredients.

Outside of school hours, Maxwell had a happy life. His aunt's house, 31 Heriot Row, had a warm atmosphere, and there he read and knitted a lot and spent hours composing letters home to his father—all the humour, bad puns and wild fantasies which should have come out in the playground emerged in those letters: signed Jas. Alex McMerkwell (an anagram), addressed to Mr. John Clerk Maxwell, Postyknowswere, (*sic*) Kirkpatrick Durham, Dumfries, and written in mirrorwriting. His cousin Jemima's artistic influence is visible in the illustrations with which he adorned some of the letters, unfortunately her talent is not. Here is one of his letters:

My Dear Mr. Maxwell—I saw your son to-day, when he told me that you could not make out his riddles. Now, if you mean the Greek jokes, I have another for you. A simpleton wishing to swim was nearly drowned. As soon as he got out he swore that he would never touch water till he had learned to swim; but if you mean the curious letters on the last page, they are at Glenlair.—Your Aff. Nephew,

James Clerk Maxwell [11]

Holidays were spent in Glenlair, tubbing in an ever more adventurous manner on the Urr, and his father often 'arranged' to be in Edinburgh to take James out on his Saturday half-holidays. If the others had their schoolboy Philosophical society, James had his father to take him to see real science, the very latest machinery and technical gadgetry, to go geologising on the hills around Edinburgh, or to go to meetings of the Royal Society of Edinburgh, to which he started to accompany his father at the age of twelve. For the schoolboy alone in Edinburgh and his father alone in Glenlair, those joint expeditions were real treats.

After a couple of years of purgatory, even life at school began to improve. Goaded beyond endurance one day, Maxwell went beserk and so earned a grudging respect from his tormentors. He also gained the enduring friendship of Lewis Campbell. Campbell's family had moved to 27 Heriot Row, and then Maxwell had someone to talk to on his way to and from school. Campbell soon realised that Maxwell was far from stupid as his new friend's quicksilver mind leaped from topic to topic; they used to stand outside Campbell's open front door with Maxwell prattling on from subject to subject, until Campbell's mother would decide she had had enough of the draught blowing around her ankles.

In the classroom Maxwell finally saw the pattern to the classics, began to understand them, and consequently began to enjoy them and his other lessons too. The syllabus for the fifth year shows that they started to learn Euclid and the Rector himself took them for classical literature, rather than the endless 'rudiments revised' of previous years. In the overall class placings, he went from nowhere, to 11th in the fifth year, and 5th in the sixth. Now that parrot-learning would no longer entirely suffice, Maxwell's real talent in maths (and English) shone through. About that time it was also discovered that Maxwell had been labouring under the additional handicap of being shortsighted, but no-one had noticed.

When Tait realised that Maxwell was not dull, he, too, formed a close and lasting friendship with Maxwell, working through mathematical problems together, and spurring each other on with new ideas: that was to continue later, when Tait was in Edinburgh and Maxwell in London, Glenlair or Cambridge, the stimulus being administered at long range by means of halfpenny postcards, scribbled down and fired off in droves.

Maxwell, finally managed to do something about his hesitancy of speech. He started to make fun of it.

P—[Punch] says that a person † of education never puts in † hums and haws; he goes † on with his † sentence without senseless interjections.
N.B. Every † means a dead pause. [12]

Then he discovered a cure. He made a plan of the large window in the Rector's room and visualised the sentences of the lesson written on successive panes of glass. During the Rector's lessons, always taken in that room, Maxwell was able to recall rapidly the sentences by looking at the appropriate pane of glass. His only fear was that he might be moved to a different desk from which his view of the window might be more oblique.[13] In part the trick was to give his gaze something neutral on which to concentrate while trying to recall his lesson, and thus got over his nervousness in having to perform 'in the round' to his large audience of classmates, but perhaps there was more to it than this.

In his book *The Mind of a Mnemonist*, (Jonathan Cape, London (1969)), the Russian psychologist A.R. Luriya analyzed the extraordinary memory of one Solomon Veniaminovich, or 'V' as he is consistently referred to. V was a tragic figure, who literally remembered *everything* that ever happened to him. This was far from being the blessing it might seem, it prevented him from ever actually doing anything. His memories were so strong—he associated colours, tastes, sounds, feelings and whole mental playlets with sequences of nonsense syllables—that he could never forget even the most trivial of them. Nor could he ever subdue the individual memories to fit them into patterns, they were too strong to be submerged. He could remember but could not collate. After failing in a succession of careers, he became a professional memory-man in a circus.

V and the other professional mnemonists that Luriya mentions all organize their memories in a strikingly similar way. V used to imagine himself walking along some familiar street, and place the list of things he wished to memorise and their associated images in the successive windows and doorways in the street. Thus he was able not only to memorise things, but able to retrieve those memories efficiently. Fast access is as important as a large data bank. Maxwell had a good memory, his learning of Psalm 119 proves that, but the development of a visual memory, a synoptic memory which could be scanned rapidly to pull out the requisite knowledge, was a great help to him in his work. The ability used to astonish his friends:

> Maxwell, as usual, showing himself acquainted with every subject upon which
> the conversation turned. I never met a man like him. I do believe there is not a
> single subject on which he cannot talk, and talk well too, displaying always the
> most curious and out-of-the-way information.[14]

Maxwell was more than a repository of wayward information. His visual memory, and its corollary, a very strong visual, geometrical imagination, gave him a broad, synoptic overview of his subject as Tait wrote in his obituary note:

> He preferred always to have before him a geometrical or physical representa-
> tion of the problem in which he was engaged, and to take all his steps with the
> aid of this: Afterwards, when necessary, translating them into symbols. . . .
> There can be no doubt that in this habit, of constructing a mental
> representation of every problem, lay one of the chief secrets of his wonderful
> success as an investigator.[15]

With his broad panoramic view of his own work, he kept a freshness of approach to research problems, an ability to step back and admire the scenery when he had reached the frontiers of knowledge. With a visual memory, the individual items can be easily re-arranged to find new patterns, but with Maxwell, this ability was combined with another: to home in on and suggest *the* crucial experiments to test his theories.

Genius cannot be analysed. The only universal qualities are total concentration and total dedication, and Maxwell had both. His visual memory and its corollary, a visual imagination, were held in check by a strong philosophic detachment, gained in part from his later studies at the University of Edinburgh. Both factors combined to give him his ability continually to re-assess his own earlier work. Normally, however, there was a dynamic tension between the two. He thought in very visual terms, using diagrams and models, geometrically and synthetically. But he was aware of the dangers of letting his visual imagination run away with him, and so he subjected his visual ideas to rigorous analysis, philosophical and mathematical.

It was his geometrical imagination which was the first to develop. In June 1844, when the rest of the class had moved in arithmetic from proportion to decimal fractions, Maxwell wrote home to his father:

> How is a' aboot the house now our Gudeman's at home? How are herbs, shrubs, and trees doing?—cows, sheep, mares, dogs and folk? and how did Nannie like bonny Carlisle? Mrs. Robt. Cay was at the church on Sunday. I have made a tetra hedron, a dodeca hedron, and 2 more hedrons that I don't know the wright names for. How do doos and Geraniums come on.—Your most obt. servt.
> Jas. Alex. McMerkwell [16]

Maxwell was proud of his rediscovery of the perfect geometrical figures—solid figures, all of whose facets are identical, for example the pyramid and the cube. The perfect figures have fascinated mathematicians since Greek times, and the astronomer Johannes Kepler embedded them in the heart of his semi-mystical model of the solar system. This was not an achievement to be buried between doos (doves) and geraniums, and he preserved his models of them to the end of his life. Now the Cavendish Laboratory museum in Cambridge has them.

Maxwell's next achievement was more easily recognised as being substantial. His father was missing James' companionship, and organised his life to spend more and more time in Edinburgh. Father and son became regular, rather than occasional, visitors to scientific meetings, at the Royal Society and at the Society of Arts in Edinburgh. At one meeting of the Society of Arts, a local printer, D.R. Hay, propounded a theory reducing art and beauty to mathematical principles. One of the problems to which he addressed himself was how to draw a perfect oval, mathematically. Young James took up the challenge—and succeeded in finding a method!

The advantage of having a parent of independent means is shown clearly in the subsequent entries in John Clerk Maxwell's diary:

February 1846.

W.25.—Called on Mr. D.R. Hay at his house, Jordan Lane, and saw his diagrams and showed James's Ovals—Mr. Hay's are drawn with a loop on 3 pins, consequently formed of portions of ellipses.

Th.26.—Call on Prof. Forbes at the College, and see about Jas. Ovals and 3-foci figures and plurality of foci. New to Prof. Forbes, and settle to give him the theory in writing to consider.

March

M.2.—Wrote account of James's ovals for Prof. Forbes. Evening—Royal Society with James, and gave the above to Mr. Forbes.

W.4.—Went to the College at 12 and saw Prof. Forbes, about Jas. ovals. Prof. Forbes much pleased with them, investigating in books to see what has been done or known in this subject. To write to me when he has full considered the matter.

Sa.7.—Recd. note from Prof. Forbes:—

Edinburgh, 6th March 1846.

My Dear Sir—I have looked over your son's paper carefully, and I think it very ingenious,—certainly very remarkable for his years; and, I believe, substantially new. On the latter point I have referred it to my friend, Professor Kelland, for his opinion.—I remain, dear Sir, yours sincerely,

James D. Forbes.

W.11.—Recd. note from Professor Forbes:—

3 Park Place, 11th March 1846.

My Dear Sir—I am glad to find to-day, from Professor Kelland, that his opinion of your son's paper agrees with mine; namely, that it is most ingenious, most creditable to him, and, we believe, a new way of considering higher curves with reference to foci. Unfortunately these ovals appear to be curves of a very high and intractable order, so that possibly the elegant method of description may not lead to a corresponding simplicity in investigating their properites. But that is not the present point. If you wish it, I think that the simplicity and elegance of the method would entitle it to be brought before the Royal Society.—Believe me, my dear Sir, yours truly,

James D. Forbes.

J. Clerk Maxwell, Esq.

Th.12.—Called for Prof. Forbes at the College and conversed about the ovals.

M.16.—Went with James to Royal Society.

T.17.—Jas. at Prof. Forbes's House, 3 Park Place, to Tea, and to discourse on the ovals. Came home at 10. A successful visit.

T.24.—Cut out pasteboard trainers for Curves for James.

W.25.—Call at Adie's to see about Report on D.R. Hay's paper on ovals.

Th.26.—Recd. D.R. Hay's paper and machine for drawing ovals, etc.

M.30.—Called on Prof. Forbes at College and saw Mr. Adie about report on Mr. Hay's paper. Jas. ovals to be at next meeting of R.S.

It was deemed improper that a fourteen-year old scholboy should address so august an institution in person, so James' paper 'On the Description of Oval Curves, and those Having a Plurality of Foci' was read for him by Professor Forbes on Monday 6 April 1846.

The paper is important in that it introduced James to Edinburgh's scientific milieu, and particularly to James Forbes, who was to become James' teacher, adviser and friend. In itself the paper is neat but not brilliant. The standard way to draw an elipse is to stick two pins in a piece of paper, and tie the ends of a loose piece of string to the pins. Then the string is pulled taut with a pencil, and the pencil run round the paper, constrained by the string. The curve that results is an ellipse. James realised that different figures could be obtained with extra loops of string.

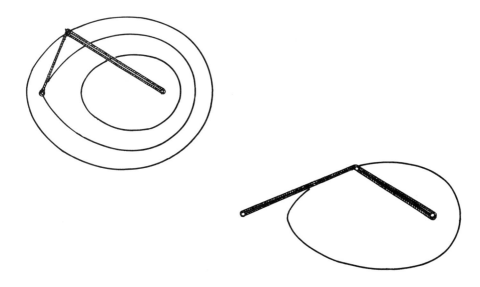

18

Technically much more advanced are some papers on the properties of the curves which he sent to Tait. He examined some of their complicated geometrical features, and showed that Descartes, one of the giants of geometry who had earlier done some work on similar curves had made mistakes.

Despite the excitement of this published piece of original research, Maxwell still had to spend another year at the Academy—he stayed on to complete the 7th year (his 6th). He finally managed to pip Lewis Campbell, who always came top overall in the exams, for the prize as best mathematician. Campbell usually came first in the maths exam, though he himself knew perfectly well that Maxwell was better:

> . . . his companions felt no doubt as to his vast superiority from the first. He seemed to be at the heart of the subject when they were only at the boundary . . .[19]

But Maxwell never regretted the generality of his education, nor the time spent on the classics. As Campbell recorded:

> And when he left, although still younger than his competitors by about a twelvemonth, he was not only first in mathematics and English, but came very near to being first in Latin. He had not yet "specialised" or 'bifurcated," although the bent of his genius was manifest. Nor have I ever heard him wish that it had been otherwise. On the contrary, he has repeatedly said to me in

later years that to make out the meaning of an author with no help excepting grammar and dictionary (which was our case) is one of the best means for training the mind.[20]

CHAPTER 5

Maxwell left the Academy at the same time as his friends, Campbell and Tait, but whereas they each spent a single year in a Scottish university before heading south to Oxbridge, Maxwell served his full term of three years at Edinburgh.

Part of the reason was his concern for his father. Cambridge was far away, and once there, the two would see each other only during the holidays. Another factor was his uncertainty about the future. While it would be pleasant to be a scientist, academic jobs in science were few and far between: there was only one professorship, occupied for life, at each of the few universities in Britain. To get one of those one had to be good, but also lucky enough that one of the sitting tenants died conveniently. Maxwell obviously was good, though how good, even he could not yet be sure, but to base his future plans on a professor's propitious demise was a big gamble.

There was also no set structure for a scientific career. In Cambridge one could become a bachelor fellow of a college, but then one had seven years in which to decide to avow permanent celibacy and become a fellow, or to find another career. In Scotland, there was not even that prop; the only course open was to become a school teacher or a private tutor and wait for something to turn up. So Maxwell considered following his father into the law and this meant studying at a Scottish university.

It was not an unusual step to take. Being an advocate gave one status and as much leisure as one chose to take, though the more taken, the less money was earned. But one could survive, and the Inns of Court harboured many an eccentric devoting his time to thoroughly non-legal pursuits. When he had graduated from Cambridge, William Thomson sounded out the Inns of Court before taking the plunge into the scientific career that was to bring him a peerage. On the other hand, Arthur Cayley (1821-1895) initially refused to take the risk. He did become a barrister, but only in the mornings. In the afternoons he was a mathematician and still found time to become one of the world's most prolific—it is a fair bet that not a single edition of his complete works, 967 papers filling 13 quarto volumes, has had all its pages cut yet. Eventually Cayley did become an academic, as the Sadlerian professor in Cambridge. To become a full-time professor instead of a part-time convey-ancer he had to accept a considerable drop in his income.

Maxwell remained in Edinburgh, undecided about his future. Nonetheless, he must have felt himself becalmed in a backwater, while his schoolfriends were progressing up the glittering ladder to fame and fortune in the South. 'The tone of his correspondence shows that he felt the disadvantages of solitude', says Campbell; but Maxwell's attitude was not helped by his friend's letters, full of the joys of Balliol, and they drew a stinging retort.

> Why do you think that I can endure nothing but Mathematics and Logic, the only things I have plenty of? and why do you presuppose my acquaintance with your preceptors, professors, tutors, etc.? . . .
>
> I don't wonder at your failing to take interest in the exponential theorem, seeing I dislike it, although I know the use and meaning of it. But I never would have, unless Kelland had explained it. . . .[1]

There is a mixture here of stiffness and self-defensive pride about the quality of Edinburgh, where Phillip Kelland was Professor of Mathematics.

But such anger was rare. The general note is one of gentle self depreciation.

> Pray excuse this wickedly perplexed letter as an effect of the paucity of our communications. If you would sharpen me a little it would be acceptable, but when there is nobody to speak to one [loses] the gift of speech. . . .[2]

Although Maxwell did not realise it at the time, his sojourn in Edinburgh provided a vital element in his scientific 'style', the characteristic way in which he set about a scientific problem, almost as individual and recognisable to him as his handwriting would be to a graphologist. This element was a philosophical detachment and caution, and it derived in part from the obligatory study of philosophy in the Scottish university degree-course.

In Cambridge, he would have received a three year cram in the rapid regurgitation of definitions from textbooks, and the mechanical solution of problems. By contrast, the Edinburgh course was a relatively leisurely affair, with no competitive exam, based on a core of philosophy. From about the middle of the eighteenth century, Scottish university students had been required to take two years of philosophy as a central feature of their course, whatever the special options they chose.

In the eighteenth century, philosophy was held in such high regard, that even one of the great scientists of all time, James Hutton, belittled his own achievements in comparison with that of the philosophers:

> Philosophy surely is the ultimate of human knowledge, or the object at which all science properly should aim; every step in science, therefore, ought to be valued, in some measure, as it tends to bring about that end. Science, no doubt, promotes the arts of life; and it is natural for human wisdom to promote these arts. But, what are all the arts of life, or all the enjoyments of the animal nature, compared with the art of human happiness—an art which is only to be attained by education, and which is only brought to perfection by philosophy! . . . The truths of science are to philosophy what the beams of wood, and the hewn stones, are to a building.[3]

Of course, science was then still in a fairly basic state, and the division between science and philosophy in many subjects was undefined.

The dominant philosophy in Scotland then was the Scottish Common Sense school. It had originated to counter the scepticism of David Hume, who had cast doubt on everything, whether one could trust one's senses, whether there is a real world outside one's skull, and worst of all, whether God exists.

The Common Sense school took it as axiomatic that the world really does exist, but granted this not insignificant premise, they wanted to make the rest of their arguments as watertight as possible, to be proof against the scathing criticism of Hume and his followers. Science, and particularly Newton's mechanics offered them their best model for an indubitably, incontrovertibly correct logical system, and so they aimed to make philosophy 'scientific'.

They had, in fact, been duped by one of the cruellest jokes in the history of science, that played by Newton on his followers when he said (in Latin) 'I do not feign hypotheses'. His rival, Robert Hooke, had casually suggested that there was some force holding the planets in orbit round the sun, and was angry that Newton, in the *Principia* (published in 1686, but reporting work done twenty years earlier), gave him no credit for the suggestion. Newton replied that it was easy enough to generate ideas at random, and some would inevitably, eventually hit targets. He had no time for such random hypothesising, but had derived the existence of an inverse square law of attraction directly from Kepler's laws, which were the synthesis of direct experimental observation. Hence *Hypotheses non fingo*.

In fact Newton was an extremely imaginative physicist. He later hypothesised a possible mechanism for gravity, and a particle model for light, which slowed down the development of optics for a hundred years. The Common Sense school initially invested Newton's statement with papal infallibility: the only way for science to work was to derive scientific laws directly from proven experimental fact. This was Induction, the model originally proposed for science by the philosopher, Francis Bacon. Thus the founder of Common Sense philosophy (and in many ways its greatest figure), Thomas Reid, could write:

> It is genius, and not the want of it, that adulterates philosophy and fills it with false theory. A creative imagination disdains the mean offices of digging for a foundation, of removing rubbish, and carrying materials; leaving these servile employments to the drudges in science, it plans a design and raises a fabric. Invention supplies materials where they are wanting, and fancy adds colouring and every befitting ornament. The work pleases the eye, and wants nothing but solidity and a good foundation.[4]

Imagination is bad, and the only true path to scientific salvation is induction. The worst thing of all was to presume to understand anything about the causes of events, only effects were observable:

> Newton, more enlightened on this point has taught us to acquiesce in a law of nature according to which the effect is produced, as the utmost that natural philosophy can reach, leaving what can be known of the agent or efficient cause to metaphysics or natural theology. This I look upon as one of the great discoveries of Newton; for I know of none that went before him in it. It has new-modelled our notion of physical causes.[5]

Thereby he ignored the fact that Newton *had* made an attempt to explain gravity. Although induction remained the holy grail to which the Common

Sense philosophers should ultimately aspire, the next century saw a steady erosion of the rigour of Reid's dogma, in particular it was realised that 'working hypotheses' were almost impossible for scientists to do without.

The danger which Reid had pointed out, of building whole theoretical edifices on false analogies, remains equally valid today, but his regime is too strict to follow literally. His follower, Dugald Stewart, recommended devising lots of analogies for the same set of phenomena, so as not to be swept away by enthusiasm for just one:

> No one has hit on the only effectual remedy against this inconvenience—to vary, from time to time, the metaphors we employ, so as to prevent any one of them from acquiring an undue ascendant over the others, either in our minds or in those of our readers. It is by the exclusive use of some favorite figure, that careless thinkers are gradually led to mistake a simile or distant analogy for a legitimate theory.[6]

There is an interesting parallel argument in Maxwell's later writings.

Thomas Brown, Stewart's successor as Professor of Moral Philosophy in Edinburgh reconciled himself to the idea of 'cause' by saying that human memory is weak:

> If phenomena were connected in our mind merely by the order of time, in which they occurred, few would be remembered nor, though memory were tenacious, could much aid be derived from it . . .
> We cannot observe the various appearances of nature, without remarking certain circumstances in which they agree; and to remark these circumstances is to arrange the similar appearances. It is thus impossible not to systematize; and hence, the question should not be whether systems be useful, but to what extent, and in what mode, they can be most usefully formed.[7]

The idea of a causal relationship can therefore be a useful mnemonic.

The very centrality of philosophy to university education, and the closeness of society in Scotland, meant that the philosophers could not avoid coming into contact with practising scientists, who might not have been so reverent in their attitudes as Hutton. Dugald Stewart was a member of the elite Oyster Club a select dining group within the ranks of the Royal Society of Edinburgh whose members included men such as John Robison, John Playfair (populariser of Huttonian geology) and Joseph Black (discoverer of carbon dioxide, and the non-inductive existence of 'latent heat' in melting and evaporation). It probably would not have taken too many glasses of claret for the other members of the club to let Stewart know what they felt about philosophers trying to dictate to them proper scientific procedure. Perhaps this accounts for the moderation of the views of the philosophers.

It is interesting to compare the teaching received by James Forbes as an undergraduate with his later views. In an essay for Professor John Wilson (the same Wilson, or 'Christopher North', who was to teach Maxwell Moral Philosophy) entitled 'On the Inductive Philosophy of Bacon his Genius and Atchievements'[8] Forbes wrote:

> [(induction)] is the only opening through which Natural Philosophy can ever arrive at firm conclusions.[9]

The prize-winning young student had merely repeated received opinion. David Brewster, a leading researcher in optics told him a different story: 'Forget all that you have heard of Lord Bacon's Philosophy. Give full rein to your imagination. Form hypotheses without number.'[10] Forbes never went quite so far as that, but his own scientific work certainly took him a long way from his youthful views.

> Is the Art of Discovery capable of being reduced to rule? One very great man, at all events, thought it was. Francis Bacon devoted the most celebrated and important of his writings to define and explode the errors by which the increase of knowledge was in his time retarded, and to systematise a positive method of discovery. In the former part of his task he was, to a great extent, successful; in the latter his failure was conspicuous. Not only did he himself not succeed in any model-investigation, but the procedure he recommended was not followed by any natural philosopher.[11]

The net result of that interplay between philosophy and science was a body of teaching that did, indeed, have a lot of Common Sense about it.

Analogy works if it is not taken to extremes, and because it can never be certain it is important to check it frequently. Thomas Brown had introduced the idea that the falsifiability of a hypothesis is its crucial aspect about a century before Popper. When choosing analogies, Stewart emphasised the importance of economy of thought, that the simplest underlying mechanism should be chosen, which, preferably should be capable itself of being independently tested.

The final figure in this succession of Common Sense philosophers was Sir William Hamilton, who also taught Maxwell. He had studied at Oxford and in Germany, and brought Aristotelian and Kantian elements to the Common Sense tradition. He adopted and put special emphasis on a point first made by Brown:

> The impression which a seal leaves on melted wax depends not on the qualities of the wax alone, or of the seal alone, but on the softness of the one and the form of the other. Change the external object which affects the mind in any case, and we all know that the affection of the mind will be different. There is no physical science, therefore, in which the laws of mind are not to be considered together with the laws of matter; and a change in either set of laws would equally produce a change in the nature of the science itself.[12]

All that one learns about the world outside is gained through the senses and thus is liable to distortion; knowledge is relative. Causal and analogical thinking are necessary consequences of the *limitations* of the brain, and therefore Occam's razor had a special place in Hamilton's Philosophy: the minimal set of assumptions necessary to explain any phenomenon are the best. There is once more a tremendous emphasis on the care necessary to approach a scientific theory.

All this had its impact on Maxwell. From the training in philosophy he received at Edinburgh, he learned the fragility of even the most apparently successful scientific theories, and that any model or analogy should be pared away to determine which parts of it are crucial and which dross.

Most scientists become devoted converts to their own theories. The mental effort involved in ploughing through uncharted scientific territory is so great that it locks the mind onto the random furrow it first followed in deriving a new result. The ability to step back from one's work, and reconsider it coolly and rationally is very rare. Part of Maxwell's talent lay in his ability to arrive at the frontiers of knowledge unhampered by too many preconceptions, leaving him fresh and willing to explore the virgin scientific territory. This was a function of his inherent genius. But the philosophic caution he learned at Edinburgh must have made him aware that frontier territory is also dangerous territory to explore unprepared and unaware.

> When a man thinks he has enough of evidence for some notion of his he sometimes refuses to listen to any additional evidence *pro* or *con*, saying, "It is a settled question, *probatis probata*; it needs no evidence; it is certain." This is knowledge as distinguished from faith. He says, "I do not believe; I know." "If any man thinketh that he knoweth, he knoweth yet nothing as he ought to know." This knowledge is a shutting of one's ears to all arguments.[13]

As he wrote to Campbell, towards the end of his undergraduate career.

This should be compared with the attitude of his friend Tait, who after the Cambridge mathematical rather than the Edinburgh philosophical training, waded boldly into the age-of-the-earth debate, brushing aside the arguments of geologists and evolutionists to place his full support behind Kelvin's apparently sound mathematical calculation of the earth's age as 100 million years:

> Let us then hear no more nonsense about the interference of mathematicians in matters with which they have no concern; rather let them be lauded for condescending from their proud preeminence to help out of a rut the too ponderous waggon of some scientific brother.[14]

Maxwell's caution permeated his work, even in his most "hypothetical" papers on electromagnetism, he found time for a *caveat*:

> If . . . we adopt a physical hypotheses, we see the phenomena only through a medium, and are liable to that blindness to faults and rashness in assumption which a partial explanation encourages. We must therefore discover some method of investigation which allows the mind at every step to lay hold of a clear physical conception without being committed to any theory founded on the physical science from which that conception is borrowed.[15]

His training had sunk deep. Here is a passage from his Presidential Address to the British Association in 1870:

> For the sake of these different types [(of mind)], scientific truth should be presented in different forms, and should be regarded as equally scientific, whether it appears in the robust form of vivid colouring of a physical illustration, or in the tenuity and paleness of a symbolical expression.[16]

Pure Stewart.

Together with this philosophical awareness of the fallibility of science went an interest in the psychology of discovery and inspiration. The actual processes of the brain are a long way off the ideal of induction:

> The machine when tried went a bit and stuck; and I did not find out the impediment till I had dreamt over it properly, which I consider the best mode of

resolving difficulties of a particular kind, which may be found out by thought, or especially by the laws of association.[17]

The views expressed in this Edinburgh letter to Campbell were later amplified:

> I believe there is a department of the mind conducted independent of consciousness, where things are fermented and decocted, so that when they are run off they become clear.[18]

He always said it took time for any ideas to percolate through the 'crust' round his brain. Despite this interest in the mechanics of the mind, Maxwell left no account anywhere of how he had himself arrived at any of his great ideas, other than what can be deduced from the form of his final papers. Partly this was proper Victorian modesty, partly perhaps because there never was any single flash of inspiration, but a steady accumulation of hunches and ideas, which he tinkered with until reaching finally a satisfactory synthesis.

The other great advantage of being an undergraduate at Edinburgh was the leisure time it bestowed on Maxwell.

> As you say, sir, I have no idle time. I look over notes and such like till 9.35, then I go to Coll., and I always go one way and cross streets at the same places; then at 10 comes Kelland. He is telling us about arithmetic and how the common rules are the best. At 11 there is Forbes, who has now finished introduction and properties of bodies, and is beginning Mechanics in earnest. Then at 12, if it is fine, I perambulate the Meadows; if not, I go to the Library and do references. At 1 go to Logic. Sir W. reads the first ½ of his lecture, and commits the rest to his man, but reserves to himself the right of making remarks. To-day was examination day, and there was no lecture. At 2 I go home and receive interim aliment, and do the needful in the way of business. Then I extend notes, and read text-books, which are Kelland's *Algebra* and Potter's *Mechanics*. The latter is very trigonometrical, but not deep; and the Trig. is not needed. I intend to read a few Greek and Latin beside. What books are you doing?
>
> ... In Logic we sit in seats lettered according to name, and Sir W. takes and puts his hand into a jam pig full of metal letters (very classical), and pulls one out and examines the bench of the letter. The Logic lectures are far the most solid and take most notes.[19]

Despite the ironical beginning, Maxwell clearly had plenty of time to pursue his private projects. He was also spurred by the knowledge that his schoolfriends were already in the glamourous South, not to waste that time.

Also noteworthy is that he was impressed by Hamilton's lectures, as well as Hamilton's thoroughly sound device of ensuring that students paid attention. Hamilton was, by all accounts, an inspiring teacher. Perhaps uniquely for a philosopher, Hamilton used to enliven his lectures with practical demonstrations.

In one of these he demonstrated the theory of colours: he painted the colours of the rainbow on segments of a wheel, and by then spinning it, gave the impression that it was white. This is particularly fascinating in that Forbes and Maxwell later did important research on colour vision using a spinning wheel. Vision is one of the senses which mediate and therefore obscure our knowledge of Nature, according to Hamilton. Our sense of spatial dimension, which derives from touch, was put on a higher pedestal by the Common Sense school than our visual sense of dimension. This can be more easily tricked: it needs the theory of perspective to lock it to tactile measurement. Our sense of colour is even more dubious, they thought. Maxwell showed by his research that, in fact, the eye is a remarkably good and consistent measurer of colour; people who are not colour blind agree amazingly well with each other on colour measurement—Nature has standardised eyes to a remarkable extent.

More than merely delivering lectures, though, Hamilton also encouraged his students to come to his house in the evenings, to discuss, borrow books from his own library, and to feel at home in the congenial atmosphere there. Maxwell had an extra claim to Hamilton's attention, not only had he been a classmate at school with one of Hamilton's sons, but his uncle, John Cay—his father's collaborator on inventions—was a close friend of Hamilton.

James Forbes, the Professor of Natural Philosophy, also encouraged Maxwell to come and visit. Having made his acquaintance over Maxwell's schoolboy paper on curves, and recognised his talent, Forbes gave the young student the run of his own private laboratory to perform his own experiments. He assisted in Forbes' experiments on colour vision (where they showed that the mixture of blue and yellow light gives a pinkish white, not green) and also did some original research of his own. While at Edinburgh he published two more papers, one, another piece of geometry, but the other a substantial piece of physics 'On the Equilibrium of Elastic Solids'. Here he developed a theory of induced double refraction in a strained solid, how a transparent solid under stress reveals that stess in rainbow coloured fringes when viewed with polarized light. This effect is now used to study stress patterns, and therefore the mechanism of possible eventual failure in models of complicated engineering structures.

The run of Forbes' laboratory was a considerable privilege. There was no tuition in experimental science in those days, nowhere for a young student to learn how to perform experiments. To provide that space and that equipment, but not to stifle Maxwell's imagination by prescribing dull experiments was one of Forbes' great contributions to Maxwell's education.

Maxwell had already had a great stroke of fortune here. His uncle John Cay had taken him to visit another eminent scientist, William Nicol, while he was still at school (everyone knew everyone else in Edinburgh).

Nicol had invented the 'Nicol prism', a device for analysing polarized light. When Maxwell went home he built himself a crude polarizer, did some experiments and recorded the results in water colours, which he sent to Nicol. He was rewarded with the gift of a pair of Nicol prisms, which he used in his later improved experiments. Typically Maxwell always kept close guard of the prisms. They are now in the Cavendish museum, though they had a close escape during one long vacation while Maxwell was at Cambridge. An overenthusiastic bed-maker thought they were rubbish and carried them off—and died before the beginning of the next term. It was only by a 'very diligent search' that they were found among the bed-maker's effects, doomed for imminent destruction.

Professor Everett has pointed out that Maxwell recognized the debt he owed to Nicol and especially Forbes in a passage, surely autobiographical, in a book review he wrote for *Nature*.

> If a child has any latent capacity for the study of nature, a visit to a real man of science at work in his laboratory may be a turning point in his life. He may not understand a word of what the man of science says to explain his operations; but he sees the operations themselves, and the pains and patience which are bestowed on them; and when they fail he sees how the man of science instead of getting angry, searches for the cause of failure among the conditions of the operation.[20]

Hamilton and Forbes may have been the twin godmothers of Maxwell's scientific career, but they were themselves implacable enemies. Hamilton diverged from the Common-Sense tradition in his views on mathematics. They had produced a pragmatic philosophy of science, based on the view that mathematics is the nearest thing to perfection that man had devised, but Hamilton disagreed. He thought philosophy was an end in itself, and that maths, which is soluble only when the number of variables is small, is not adapted to the real problems of life. Philosophy, though, is central to man's existence, and should be central to a university education.

Forbes, however, felt that he was wasting too much of his time teaching elementary maths to his students: They were not being taught enough at school because they were being fitted for a general education at university. As the body of scientific knowledge was advancing, he wanted to see specialist scientific degrees at Edinburgh, as, he noted, Cambridge had recently introduced. He wrote to William Whewell in Cambridge.

> The state of preparation here is low to a degree which, with your high academic notions, fostered by the spirit of your noble University, must appear to you almost incredible.[21]

The row between the two men exploded over the appointment of a new professor of maths in 1838. Forbes agitated to get Phillip Kelland brought from Cambridge, to teach calculus, an understanding of which is vital for science, generally formulated in terms of differential equations.

Hamilton on the other hand, thought geometry was still the most important branch of mathematics to learn.

> The mathematical process in the symbolical method [i.e., the algebraical] is like running a railroad through a tunnelled mountain, that in the ostensive [i.e. the geometrical] like crossing the mountain on foot. The former causes us, by a short and easy transit, to our destined point, but in miasma, darkness and torpidity, whereas the latter allows us to reach it only after time and trouble, but feasting us at each turn with glances of the earth and of the heavens, while we inhale the pleasant breeze, and gather new strength at every effort we put forth.[22]

The algebraic approach was like a dark tunnel because an answer could often be derived by a programmatic manipulation of symbols without the manipulator needing to understand what he was doing. Indeed in algebra, the use of imaginary numbers is based on the symbol i, which represents the square root of −1. For the inductivist Common Sense school this did not even *exist* in any real sense of the word. They called imaginary numbers Impossible Quantities. Even where the symbol *i* dropped out of the final, concrete answer, intermediate steps involving *i* could not be individually meaningful. Similarly they saw calculus as based on the division of infinitesimally small numbers by other equally infinitesimal quantities, which is not a well-defined mathematical operation; in both calculus and complex algebra (algebra involving *i*) the distinction between legitimate operations, giving meaningful answers, and illegitimate manipulations giving rise to absurdities, was not always satisfactorily delineated for them, because they could not attach concrete meaning to each intermediate step. John Playfair wrote:

> The geometer is never permitted to reason about the relation of things which do not exist . . . the analyst continues to reason about the characters after nothing is left which they can possibly express.[23]

and he was defending complex analysis. Henry Brougham and John Leslie were stronger:

> Mathematicians have been more attentive to improve and extend their methods, than solicitous to examine the principles on which they are founded. Men of a scientific turn, who wish to reason as well as to compute, and who will not assent to the truth of the conclusion without fully comprehending every step in the reasoning that leads to it, have justly to complain of the mystery and paradox attending the use of impossible quantities.[24]

> It is indeed the reproach of modern analysis to be clothed in such loose and figurative language, which has created mysticism, paradox, and misconception. The algebraist, confident in the accuracy of his results, whenever they become significant, hastens through the successive steps to a conclusion without stopping to mark the conditions and restrictions implicated in the problem.[25]

This perennial attack on complex numbers was even absorbed by Maxwell and regurgitated in one of the few really crass statements he ever made.

> By an imaginary surface [(used for cutting Faraday's lines of Induction)] is meant a surface which has no physical existence, but which may be imagined to exist in space without interfering with the physical properties of the substance which occupies that space. Thus we imagine a vertical plane dividing a man's head longitudinally into two equal parts, and by means of this imaginary surface we may render our ideas of the form of his head more precise, though any attempt to convert this imaginary surface into a physical one would be criminal. Imaginary quantities, such as are mentioned in treatises on analytical geometry, have no place in physical science.[26]

The whole of twentieth century physics depends on the use of *i*.

In hindsight, though, both Hamilton and the Common Sense school, with their devotion to geometry, and Kelland wanting to teach calculus were correct.

Calculus *is* vital for science, but for a generalist, no branch offers a better training in how to think mathematically than geometry. It enforces clear, constructive, logical, rigorous thought; the ground rules are simple and there are no mechanical programmes which will churn out results. It is much to be regretted that all so-called new maths syllabuses have decided that geometry is too old fashioned to be taught properly, it is still the best basic training in mathematical thinking.

In this instance, however, Forbes won, mostly through political adroitness, and Kelland was appointed rather than Hamilton's candidate for the job, David Gregory. Ironically, almost the first thing that happened to Kelland on his arrival was that the 16-year old William Thomson spotted a serious error in Kelland's work, and moreover in calculus. Kelland had accused Thomson's hero, the French mathematical physicist Fourier, of making a mistake in his classic book on the theory of heat. Thomson realised that, in fact, it was Kelland who had made the mistake, wrote an angry paper and sent it to David Gregory. Gregory courteously passed it to his rival, who wrote back an even angrier reply. Thomson and his father went back over the paper, removed all passages liable to offend, and returned it to Kelland, who now pronounced himself charmed with the revised paper and had it published. Thomson's first publication.

Maxwell's school mathematics had been geometrical, and he always favoured the geometrical approach, as a Cambridge colleague vividly remembered.

> I remember one day in lecture, our lecturer had filled the blackboard three times with the investigation of some hard problem, and was not at the end of it, when Maxwell came up with a question whether it would not come out geometrically, and showed how with a figure, and in a few lines, there was the solution at once.[27]

But this is Maxwell's pictorial imagination as much as anything (the quote from Hamilton above is very apposite to Maxwell's own case). When Maxwell wanted to, he proved himself perfectly adept at high-powered analysis.*

Indeed it was at Edinburgh through his private reading, that Maxwell's mathematical knowledge branched out. Tait noted that the university library's records showed that Maxwell had borrowed Fourier's *Theorie de la Chaleur*, Monge's *Geometrie Descriptive*, Newton's *Optics*, Willis's *Principle of Mechanism*, Cauchy's *Calcul Différentiel* and Taylor's *Scientific Memoirs*. He also read Boole's *Mathematical Analysis of Logic* and Poisson's *Mechanics*, amongst others. Like Thomson before him, Maxwell was bowled over by Fourier's book, calling it a 'great mathematical poem'—and actually bought his own copy, for 25 shillings, a vast sum in those days.[28]

If Maxwell had ever been called upon to specify his position in the Forbes-Hamilton contest, he would probably have sided with Hamilton. He was a product of a generalist school education and felt its benefits, as Campbell noted:

> (At school) he had not yet "specialised" or "bifurcated," although the bent of his genius was manifest. Nor have I ever heard him wish that it had been

---

* Pierre Duhem in his seminal book on the history of science *The Aim and Structure of Physical Theory* denigrates British science of the 19th century (and even more its slavish European adherents) as being rigidly mechanical and geometrical in outlook, and sloppy. It is true that Maxwell and Thomson did indeed *like* to visualise their theories—but both did plenty of important analytical work to show the opposite, rigorously mathematical sides to their natures.

otherwise. On the contrary, he has repeatedly said to me in later years that to make out the meaning of an author with no help excepting grammar and dictionary (which was our case) is one of the best means for training the mind.[29]

He was as good as his word, and used in later life to translate odd passages of Latin for the mental stimulus it provided, much as some people tackle crosswords. At university he enjoyed his philosophy, and geometry always remained his favourite branch of mathematics, and both were useful to him as a physicist.

Thus the traditional general Scottish education had done him no harm, and he would have had difficulty in seeing it might prejudice the education of others. He had the somewhat patrician attitude towards science of the dedicated amateur, who can afford to devote more of his time to his subject than can the professional. For specific scientific training he believed in close contact with nature and in training the senses to be alert, but was unable to see that others without his country upbringing might not have developed such powers, or that they could be taught.

He was never a dynamic leader of a research team, but rather gave students the opportunity to follow their own ideas and develop their own insights, as he developed his. As a way for not spoiling native genius, Maxwell's method is fine, it allows it to develop at its own rate. As a recipe for producing the competent scientists that the age of scientific industrialisation was going to demand, it would have been a disaster.

Maxwell's gentleman-amateur attitude to science is shown nowhere better than in his letters to Campbell during his holidays from Edinburgh. For him, holidays were time when, freed of all constraints, he could work twice as hard at things which interested him. In the summer of 1848 he took up electroplating as a hobby, covering everything in sight with layers of copper in his cubby hole laboratory at Glenlair. In September he gave Campbell this account of his breathtaking Victorian industry.

> Glenlair, 22nd Sept.1848.
>
> . . . When I waken I do so either at 5.45 or 9.15, but I now prefer the early hour, as I take the most of my violent exercise at that time, and thus am *saddened down*, so that I can do as much still work afterwards as is requisite, whereas if I was to sit still in the morning I would be yawning all day. So I get up and see what kind of day it is, and what field works are to be done; then I catch the pony and bring up the water barrel. . . . Then I take the dogs out, and then look round the garden for fruit and seeds, and paddle about till breakfast time; after that take up Cicero and see if I can understand him. If so, I read till I stick; if not, I set to Xen. or Herodt. Then I do props, chiefly on rolling curves, on which subject I have got a great problem divided into Orders, Genera, Species, Varieties, etc.
>
> One curve rolls on another, and with a particular point traces out a third curve on the plane of the first, then the problem is:—. . . . I have proved that the equi-angular spiral possesses the property, and that no other curve does. This is the most reproductive curve of any. I think John Bernoulli had it on his tombstone, with the motto *Eadem mutata resurgo*. There are a great many curious properties of curves connected with rolling. Thus, for example . . .
>
> After props come optics, and principally polarised light.

Do you remember our visit to Mr. Nicol? I have got plenty of unannealed glass of different shapes, for I find window glass will do very well made up in bundles. I cut out triangles, squares, etc., with a diamond, about 8 or 9 of a kind, and take them to the kitchen, and put them on a piece of iron in the fire one by one. When the bit is red hot, I drop it into a plate of iron sparks to cool, and so on till all are done. I have got all figures up to nonagons, triangles of all kinds, and irregular chips. . . .

I have got a lucifer match box fitted up for polarising, thus. The rays suffer two reflections at the polarising angle from glasses A and B.

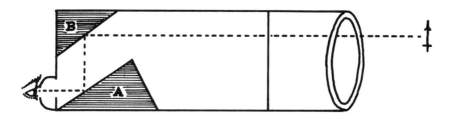

[Here follow thirteen diagrams of patterns in triangles, squares, pentagons, and hexagons.] These are a few of the figures one sees in unannealed glass.

Pray write soon and tell when, how, and where by, you intend to come, I suppose when you come I will have to give up all my things of my own devising, and take Poisson, for the time is short, and I am very nearly unprepared in actual reading, though a great deal more able to read it.

I hope not to write any more letters till you come. I seal with an electrotype of the young of the ephemera. So, sir, I was, etc.[30]

A summer walking tour with Professor Forbes and a select party, included carrying stones up Ben Nevis to build a 3-foot cairn on the top, and for fun, the calculation of the time it would take a log to slide under gravity from the cairn to Fort William, following a curve of a particularly revolting shape (Maxwell reckoned 49.6 seconds).

Forbes, outside of his arguments with Hamilton, seems to have been a pleasant and agreeable man, who managed to combine his profession and interests with remarkable skill. He is best-known for his invention of the seismometer, and the discovery that glaciers moved viscously (like sluggish rivers of ice). These geophysical interests enabled him to get off into mountainous country, to go walking and climbing as much as doing science. He was called 'The Discoverer of the Alps', for his climbs there, made while studying glaciers. In Scotland, he made the first ever ascent of Sgurr nan Gillean, in the Cuillins on Skye.

In the time he spent in Forbes' laboratory Maxwell matured and decided that science and not the law was to be his career. He also gained an experimental competence, and an appreciation of the role of experiment in physics. Forbes had an intense dislike of the mathematical physicist's game, of inventing mathematical models because they are soluble rather than because they have any relevance to the physical problem in hand.

Forbes also taught him the importance of clarity and accuracy of expression. While at Edinburgh he wrote a paper on Elasticity, the original draft of which was returned to him by Forbes.

> Edinburgh, 4th May 1850.

> My Dear Sir—Professor Kelland, to whom your paper was referred by the Council R.S., reports favourably upon it, but complains of the great obscurity of several parts, owing to the abrupt transitions and want of distinction between what is assumed and what is proved in various passages, which he has marked in pencil, and which I trust that you will use your utmost effort to make plain and intelligible. It is perfectly evident that it must be useless to publish a paper for the use of scientific readers generally, the steps of which cannot, in many places, be followed by so expert an algebraist as Prof. Kelland;—if, indeed they be *steps* at all and not assumptions of theorems from other writers who are not quoted.[31]

Maxwell took Forbes' comments to heart, and his subsequent papers are a model for the scrupulousness with which he bestowed credit on earlier writers, and for the clarity and resonance of his prose-style. Professor Charles Coulson, biassed though he may have been, said:

> It has long been one of my views that the best scientists are often the most attractive writers. Among many names that I could list, . . . I should choose Maxwell as one of the best. In any anthology of English prose writing of the Victorian Age, I should most assuredly include some contribution of his.[32]

Now that he was set on a scientific career, it was immediately obvious that Maxwell had to gain the academic cachet of a good Cambridge degree. The maturity he had found in Edinburgh, from contact with Forbes and Hamilton is shown nowhere better than in this passage from a letter to Campbell. He is at the point of departure for Cambridge, and lists the things he must mug up for their exams:

> 3. Nat. Phil.—Simple mechanical problems to produce that knack of solving problems which Prof. Forbes has taught me to despise. Common Optics at length; and for experimental philosophy, twisting and bending certain glass and metal rods, making jellies, unannealed glass, and crystals, and dissecting eyes— and playing Devils.[33]

Cambridge was an unavoidable rung on the scientific ladder, but Maxwell was fully aware of the idiocies of the Cambridge system he was letting himself in for.

His final act before leaving Scotland was to attend the British Association meeting in Edinburgh, where, from the floor, he attacked the views on optics of the very eminent Sir David Brewster. Professor William Swan later recalled the incident:

> My earliest noteworthy recollection of Clerk Maxwell dates from 1850, when the British Association was in Edinburgh. One day when a paper had been read, which, if I remember rightly, was Brewster's, a young man rose to speak. This was Clerk Maxwell. His utterance then, most likely, would be somewhat spasmodic in character, as it continued to be in later times, his words coming in sudden gushes with notable pauses between; and I can well remember the half-puzzled, half-anxious, and perhaps somewhat incredulous air, with which the

president and officers of the section, along with the more conspicuous members who had chosen "the chief seats" facing the general audience, at first gazed on the raw-looking young man who, in broken accents, was addressing them. For a time I was disposed to set down his apparent embarrassment to bashfulness; and such, I daresay, was the general impression. Bashful, very likely, in some degree he was. But, at all events, he manfully stuck to his text; nor did he sit down before he had gained the respectful attention of his hearers, and had suceeded, as it seemed in saying all he meant to say.[34]

CHAPTER 6

The Cambridge system was based on the Tripos exam. To do well in the Tripos was essential for a young scientist, it was the yardstick by which talent was subsequently judged no matter what body of achievement was afterwards gained. That was true not only for science; to go on to take an honours degree in Latin or Greek, the aspiring classicist was first obliged to get a sufficient pass in the Maths Tripos. Someone with a good Tripos result was regarded as an excellent prospect to become a high-court judge or archbishop. To qualify as 'next prime-minister but three', the skilful choice of one's parents, and the possession of a stiff upper lip were crucial, but success in the Tripos was also useful. Approximately forty Church of England bishops in the first half of the nineteenth century were helped to their positions by being Wranglers—getting firsts in the Tripos.

What was this ferocious exam? It consisted of a full seven days of examinations in the winter of the fourth year at Cambridge; a grand total of two hundred questions to be answered. There was no limit set on the number that could be attempted. The first three days' exams were based mainly on standard bookwork (the books in question being Euclid and the *Principia*) and passing those exams was sufficient for most students. For those who wanted Honours degrees—in classics as well as mathematics—they had to endure four more days of more difficult problems. The very best mathematicians would finally take a further exam—for the Smith's Prize. Although this did not have the same public acclaim as the Tripos, for the professionals it rated higher, the problems were more difficult, and there was no regurgitation of bookwork, which in the early stages of the Tripos could build up an unassailable cushion of marks for a mathematical automaton.

And the course was designed to make the undergraduate just that. The lectures were optional, and the bulk of his learning was acquired from a private tutor; there would be daily group sessions with a paid freelance mathematician outside the college system who had set up to get students through the Tripos, whose very livelihood depended on the success rate he achieved, and whose methods were therefore totally exam-oriented. The student group would be set regular time-trial exams; they were taught to write quickly and accurately, and trained in the short cuts and trickery of problem solving, rather than in the fundamentals of the subject. The tutor was paid to be a slave driver, one tutor of that time was to be avoided because, 'he doesn't slang his men enough'. Maxwell's own tutor had fifteen students in his group, which drew the wry remark from Maxwell's father, 'He should get one more pupil and drive 16 in hand'.

An exception to the rule of private tutors seems to have been St John's College, which made a habit of dominating the Tripos. There, a succession of college tutors, Isaac Todhunter being the last and most celebrated, trained their teams to a clincial peak of automation. In his year, William Thomson was generally expected to be Senior Wrangler (i.e. the top first), not only was he brilliant (he had already produced major research), he was quick-witted too. But there was another contender in the field, Samuel Parkinson of St John's. Here is an account of the race:

> The Johnians were believed to be short of good men and owned it themselves. But now their best man suddenly came up with a rush like a dark horse, and having been spoken of before the Examination only as likely to be among the first six, now appeared as a candidate for the highest honours. E—was one of the first that had a suspicion of this, from noticing on the second day that he wrote with the regularity and velocity of a machine and seemed to clear everything before him. And on examining the work, he could scarcely believe that the man could have covered so much paper with ink in the time (to say nothing of the accuracy of the performance), even though he had seen it written out under his own eyes.[1]

Parkinson, who had spent the previous six months practising writing out maths problems against time to inure his wrists to cramp, did indeed come first, though Thomson gained a measure of revenge by beating him convincingly in the Smith's prize rerun.

Years later, Kelvin talked about some of this to Lord Rayleigh, and fortunately R.J. Strutt, Rayleigh's son and biographer was listening; they began by talking about Challis, who was the Plumian Professor of Astronomy at Cambridge, and who looked for but failed to find Neptune:

> *Rayleigh* "He [(Challis)] had a curious knack of getting hold of the wrong end of the stick. Do you remember what he called his "law" in hydrodynamics?
> *Kelvin* Yes. He set it when he examined me for the Smith's prize. I could not see that it was right, and I took it round to Stokes in the evening, but he knew about it before.
> *Rayleigh* Stokes knew it wasn't right?
> *Kelvin* Yes. Parkinson did it though.
> *Rayleigh* Did it according to Challis, you mean?
> *Kelvin* Yes.
> *R.J. Strutt* And I suppose scored many marks by it?
> *Kelvin* I have no doubt he did.

Rayleigh's own attitude to the Tripos is also brought out in the biography.

> The earlier papers were almost a writing race for the better men, and he [(Rayleigh)] used to tell how he had answered one question during the time that the answers were being collected from other candidates. It was an advantage to be low down in alphabetical order![3]

A further hazard in the Tripos steeplechase was presented by the weather: the exam took place in the Cambridge Senate House in early January. It is a matter of general complacence in Cambridge that there is no physical barrier in Europe to prevent their direct enjoyment of the bracing Siberian gales. Even though this may not be true geographically, it certain is true psychologically.

> The particular time of the year when the examination is held gives rise to an occasional source of failure of a rather odd sort.
> The Senate House being a large, airy, stone floor building, can be but imperfectly warmed if the weather be damp or severely cold. Thus a man with any tendency to imperfect circulation becomes chilled, especially in his hands, and with chilled hands, he is disabled to a considerable extent from writing. The first year I was at Cambridge, one of our best Trinity men, afterwards a Fellow, lost fifteen or eighteen places among the Wranglers, as he believed, and as previous and subsequent successes entitled him to believe, solely from being frozen up.[4]

Of course, not all students were potential Wranglers. Some were the sons of the wealthy, there to acquire polish, and at least a third of the undergraduates were destined for the church. The degree awarded to them was irrelevant, and they tended to be the wildest group in the University. Drinking, rowing and gambling were favourite occupations, but the most pernicious of all was billiards, 'when a young man becomes fairly engaged at billiards, he seldom does anything else very regularly'. Then there was horse racing too, and the few illusions Maxwell might have taken with him to Cambridge were rapidly dissipated

> Facts are very scarce here. There are little stories of great men for minute philosophers. Sound intelligence from Newmarket for those that put their trust in horses, and Calendristic lore for the votaries of the Senate-House. Man requires more. He finds x and y innutritious, Greek and Latin indigestible, and undergrads. nauseous. He starves while being crammed. He wants man's meat, not college pudding. Is truth nowhere but in Mathematics[5]

Things began to improve when Maxwell migrated to Trinity. He had begun at Peterhouse College (where Thomson and Tait had been before him), which was a small college with a high reputation. Because there were so many good mathematicians there already, Maxwell was advised by his college tutor that his chances of getting a fellowship to stay on after his degree would be better if he moved over to Trinity.

The advice was, in fact, excellent. One of his rivals at Peterhouse was E.J. Routh, who did become Senior Wrangler in Maxwell's year (as Tait had been two years before). Routh was no mere automaton, however, he was a very good mathematician, who went on to do fine research, and, as the Peterhouse tutor, to maintain Peterhouse as a mathematical powerhouse.

Rayleigh later wrote:

> I have always been under the impression that Routh's scientific merits were underrated. It was erroneously assumed that so much devotion to tuition could leave scope for little else.[6]

In fact Routh did important work on, amongst other things, the use of generalized co-ordinates, and Lagrangian methods—and has a mathematical quantity—the Routhian—named after him.

Trinity was the biggest of the Cambridge colleges and the most liberal in atmosphere, and within its walls Maxwell found a congenial group of undergraduates amongst whom his talents could blossom. Also, when his first year exam results became known, Maxwell was 'accepted' to become the fifteenth member of William Hopkins' Tripos team. Despite his father's joke, that was a real stroke of luck. Hopkins was the recognised Wrangler-maker at Cambridge. Year in, year out his students did well in the Tripos—though, of course, once the trend was established he was able to some extent to coast on his reputation by waiting until the end of the first year to pick from the cream of that year's crop a matched team to 'drive'. Even more remarkable than their record of Tripos successes, was the fact that so many of his students also went on to become great scientists: Stokes, Thomson, Tait, Routh, Maxwell himself, Ellis and Cayley were all Hopkins Wranglers. Despite drilling them in exam technique, Hopkins also managed *not* to stifle their real scientific abilities, and that was his real achievement.

That was probably due, in part, to his being a practising scientist himself. He had started life as a gentleman farmer, but when his wife died, he decided to go to Cambridge to study. He did well in the Tripos, was ineligible for a college fellowship, having remarried, and so stayed in Cambridge and set up as a private tutor. Nonetheless he maintained wide interests in all aspects of science, was a founder member of the British Association and he was also an original geophysicist: perhaps his early career as a farmer inspired him in that direction. He did research on the internal structure of the Earth (Thomson used some of his former tutor's results as ammunition in his fight with the geologists), the processes involved in creating mountains and the mathematical theory of glaciers. Here he came into conflict with Forbes who was concerned that simplifications would inevitably have to be made in a very complex natural phenomenon in order for a soluble mathematical description to be produced. Ultimately, he thought, an accurate theory would have to account for the individual movement of every atom in the glacier. In the best Cambridge tradition Hopkins produced a model which, though incomplete, did at least model some of the features of a glacier. Forbes would have no truck with such idle mathematical games whose danger lay in the temptation to devise a model which was mathematically tractable, rather than which represented the real physics

> It is indeed, an eminent criterion of our being well employed that our labour should not terminate in itself, but be fruitful. The solution of curious or highly abstract problems, though to a limited extent a commendable exercise, is ever to be guarded against when it is valued on its own account as a dexterous achievement. In this respect, it ranks with the dexterity of the chess-player, and no higher.[7]

At Edinburgh, Maxwell had learned to take the best elements from the teachings of the two enemies, Forbes and Hamilton. So too at Cambridge he learned to take the best from the Cambridge Applied Maths tradition without losing the intuitive feeling and flair for the subject he had developed under Forbes. The essential skill he now acquired was that of judicious approximation. To solve an equation, often certain of the mathematical terms in it have to be ignored; the skill lies in knowing which terms may be so ignored without distorting the physics. Acquire that skill Maxwell certainly did, though perhaps not in the methodical manner Hopkins would have wanted:

> He brought to Cambridge in the autumn of 1850 a mass of knowledge which was really immense for so young a man, but in a state of disorder appalling to his methodical private tutor. Though that tutor was William Hopkins, the pupil to a great extent took his own way; and it may safety be said that no high wrangler ever entered the Senate-House more imperfectly trained to produce "paying" work than did Clerk Maxwell.[8]

Tait wrote in his obituary for Maxwell. Hopkins had been Tait's tutor too. It was Hopkins whom Maxwell astonished with his geometrical skills:

> Our lecturer had filled the black board three times with the investigation of some hard problem and was not at the end of it, when Maxwell came up with a question whether it would not come out geometrically, and showed how with a figure, and in a few lines, there was the solution at once.[9]

> One striking characteristic was remarked by his contemporaries at Hopkins's lectures. Whenever the subject admitted of it he had recourse to diagrams, though the rest might solve the question more easily by a train of analysis.[10]

Hopkins soon realised that Maxwell was altogether in a class of his own. A fellow member of that Hopkins' team, W. N. Lawson, noted in his diary:

> July 15, 1853—He (Hopkins) was talking to me this evening about Maxwell. He says he is unquestionably the most extraordinary man he has met with in the whole range of his experience; he says it appears impossible for Maxwell to think incorrectly on physical subjects; that in his analysis, however, he is far more deficient; he looks upon him as a great genius, with all its eccentricities, and prophesies that one day he will shine as a light in physical science, a prophecy in which all his fellow-students strenuously unite.[11]

For Maxwell most of the Tripos work seems to have been fairly easy, and left him with enough time for his other interests; as the invaluable Lawson again remarks:

> Maxwell was, I daresay you remember, very fond of a talk upon almost anything. I well recollect how, when I had been working the night before and all the morning at Hopkin's problems with little or no result, Maxwell would come in for a gossip, and talk on while I was wishing him far away, till at last, about half an hour or so before our meeting at Hopkins's, he would say—"Well, I must go to old Hop's problems;" and by the time we met there they were all done.[12]

There was probably an element of bravado in this, as there was in the letter he wrote home in the summer of 1853, when he was gearing himself up for the actual Tripos that winter:

> Trin.Coll., 7th June 1853.
> If any one asks how I am getting on in Mathematics, say that I am busy arranging everything, so as to be able to express all distinctly, so that examiners may be satisfied now, and pupils edified hereafter. It is pleasant work, and very strengthening, but not nearly finished.[13]

Two weeks later, while staying with a friend, Maxwell collapsed. It was diagnosed as brain fever, always popular with Victorian doctors, brought on by overwork.

This happened at the home of the Rev. C.B. Taylor, and he and his family devotedly nursed Maxwell back to health. Maxwell never forgot their kindness, and the illness made him more religious than ever. It did not, however, entirely dampen his powers of observation: he drew an acute distinction between the independence he had had to learn as an only child, and the lack of individual spirit in the members of a large close-knit family group.

He recovered, returned to Cambridge and resumed work, gently at first, for the Tripos. For relaxation during the actual course of the exam, he seems to have managed to persuade a group of other candidates to join him in the evenings in doing experiments with his private laboratory stock of magnets and gutta percha.

When the results came out, he was second in the Tripos, to Routh, and equal first with him in the Smith's Prize. Everyone was delighted, except his father, for whom nothing but the first place would have done:

> I heartily congratulate you on your place in the list. I suppose it is higher than
> the speculators would have guessed, and quite as high as Hopkins reckoned on . . .
> I think I will call on Dr. Gloag to congratulate him. [14]

For Maxwell the hard grind was over. For the past three years he had done little or no research work on his own, but now he was free to follow his own interests once more, first as a bachelor-scholar then as a fellow of Trinity College.

Indeed the first pieces of research he did, seem to be infected with a spirit of gay abandon. One article was on how a sheet of paper tumbles through the air. Another paper investigates, geometrically, a focussing lens which works by having a refractive index that varies with its radius, but the same thickness throughout. It was inspired, Maxwell wrote, by investigating the eyeball of a fish—which approximates this trick. There is a college legend that this epic confrontation took place at college breakfast one morning with a cold, greasy kipper.

Another piece of research which entered college legend was Maxwell's experiments to test the ability of cats to land on their feet even though dropped on their back. As Maxwell pointed out, to his critics who branded him as a vivi-sectionist, the point is to see how quickly cats can do it, which therefore involves dropping them from as *low* a height as possible. He dropped cats on to his bed from the great height of two inches—and they still landed on their feet!

Actually, Maxwell was more interested in dogs than in cats, and outside of college, where dogs were not allowed (a rule the head of Trinity, William Whewell, rigorously enforced), Maxwell always seems to have had a shaggy dog of somewhat indeterminate breed follow him around. When he returned late to Cambridge as Professor, he once persuaded his dog Toby to sit on an insulated stand while rubbing him with a piece of cat's skin. He found that Toby became positively charged, electrostatically, though until then it had been assumed that cat's fur was positive to all other substances. Maxwell remarked that 'a live dog is better than a dead lion' and wisely left the obvious follow-up experiment of testing live dog against a *live* cat. The result of the experiment appears even today to be in doubt.

All this reveals a lively sense of humour and sociability in Maxwell, which burst into flower at Cambridge. When he arrived, he was regarded with suspicion as an eccentric. He had an aversion to formal clothing—as Campbell delightfully puts it 'His dress was . . . remarkable for the absence of everything adventitious (i.e. starch, loose collars, cufflinks etc.)'—and like many before and since, he tried varying his working and sleeping hours.

> I have been trying an experiment of sleeping after hall. Last Friday I went to bed from 5 to 9.30, and read very hard from 10 to 2, and slept again from 2.30 to 7.
>
> I intend some time to try for a week together sleeping from 5 to 1, and reading the rest of the morning. This is practical scepticism with respect to early rising.[15]

From 2 to 2.30 a.m. he took exercise, running along the college corridors. The inhabitants of rooms along his path took to sitting by their doors and showering him with boots, hair brushes and other missiles as he passed. He returned to more normal sleeping hours.

There were two special factors now in his favour. In Edinburgh his Galloway accent had instantly marked him as a rustic. To the ignorant Cambridge men, however, he was just another Scot, and the Scottish accent was then the only regional accent tolerated: it indicated hard work, canniness and a determination to succeed. A regional English accent, on the other hand, meant merely that one's parents could not afford the fees at a private school. Maxwell was also two years older, maturer and far more widely read than his contemporaries, which gave him a degree of authority. The impression he made on Lawson has already been noted (see p.62), and he equally impressed Professor Hort:

> During the time that I knew him I can recall no perceptible signs of change other than quiet growth, and suspect that he attained too early and too stable a maturity to receive easily a new direction from any kind of intercourse with his University contemporaries.[16]

and the Rev. Butler who later became Master of Trinity,

> His position among us—I speak in the presence of many who remember that time—was unique. He was the one acknowledged man of genius among the undergraduates.[17]

His abilities were recognised by an invitation to join the Apostles, a select essay club consisting of the twelve brightest undergraduates in Cambridge, or at least that is how they thought of themselves. Tennyson, G.E. Moore, Bertrand Russell, Lytton Strachey and J.M. Keynes were all members. The Apostles met on a Saturday afternoon in the rooms of one of their number, and after eating their anchovies on toast one of them would read a philosophical essay on a subject of his choosing. Each Apostle then gave his views on the topic and then a general debate ensued. Finally everyone entered their signed opinions in the Society's records.

The series of essays which Maxwell presented to the Apostles is very interesting. They are immature works with a good deal of posturing: elegant phrases tossed off for the sake of their elegance rather than their truth (perhaps it was here that Maxwell learned to write so elegantly, a feature which marks his later scientific papers), and there is a deal of deep theorising based on extremely tenuous evidence (e.g. a model of the *invariable* development of pseudoscience based on the sole example of animal magnetism). Nonetheless there are some very interesting insights buried in these essays.

In one essay 'Are There Real Analogies in Nature', Maxwell begs the question by drawing a convoluted analogy between cause and effect in Nature and in the moral sphere (inevitable retribution). He even laughingly admits it himself, 'I have been somewhat diffuse and confused on the subject of moral law'. But the essay gains its interest from its date, February 1856, for by then he had already become seriously interested in the theory of electricity and magnetism. He set himself the task of reading both the work of Michael Faraday, whose great series of experimental researches formed the basis of Maxwell's subsequent research, and also the work of William Thomson.

To begin with, Maxwell was still in the grip of the notion that the best and surest way to approach a subject was through induction. He wrote to a friend in June 1855:

> It is hard work grinding out "appropriate ideas", as Whewell calls them. However, I think they are coming out at last, and by dint of knocking them against all the facts and ½-digested theories afloat, I hope to bring them to shape, after which I hope to understand something more about inductive philosophy than I do at present.
>
> I have a project of sifting the theory of light and making everything stand upon definite experiments and definite assumptions, so that things may not be supposed to be assumptions when they are either definitions or experiments.[18]

By the following February, however, his ideas as revealed in his Apostles essay, had shifted:

> There is nothing more essential to the right understanding of things than a perception of the relations of number. Now the very first notion of number implies a previous act of intelligence. Before we can count any number of things, we must pick them out of the universe, and give each of them a fictitious unity by definition. Until we have done this, the universe of sense is neither one nor many, but indefinite. . . . The dimmed outlines of phenomenal things all merge into another unless we put on the focussing glass of theory and screw it up sometimes to one pitch of definition, and sometimes to another, so as to see down into different depths through the great millstone of the world.[19]

By trying his philosophical prejudices against the real nature of science, he had realised a major point missed or merely dismissed by the philosophers. If experimental facts make up the *body* of science, then theory is its skeleton; it gives shape to the body. Without tentative hypotheses to begin with, one just has a shapeless pile of experimental 'facts' collapsed on the ground. Without some theoretical prejudice, the fact that the speed of light of any wavelength in free space is not infinite but is actually 186,000 miles per second, and the fact that it was raining at five past eleven this morning are equally significant statements about Nature. One cannot begin to do science without some ranking of facts into categories of significance, and that requires theory. Faraday himself, as Maxwell soon realised, was no mere recorder of experimental data, but planned his research on a sophisticated model of Nature.

Nonetheless, he was enough of a philosopher to sound a cautionary note:

> Perhaps the 'book', as it has been called, of nature is regularly paged; if so, no doubt the introductory parts will explain those that follow, and the methods taught in the first chapters will be taken for granted and used as illustrations in the more advanced parts of the course; but if it is not a 'book' at all, but a magazine, nothing is more foolish to suppose that one part can throw light on another.[20]

The hypothesis that Nature is ordered underlies the whole scientific endeavour.

Maxwell's evolving personal philosophy of science is highlighted in a later essay, dated October 1856, 'Unnecessary Thought' where one of the points he makes most forcibly is that abstract thought needs to be continually checked through contact with reality, and it is more important that our thoughts should have a living root in experience than that they should be perfectly self-consistent at any particular stage of their development. This is an important statement, it is at once a rejection of induction and also of the extremes of the Cambridge tradition:

> In Natural Philosophy there are a great many different things which must be made our own, before we can have right notions upon what is to follow, and we have this great advantage over the students of many other sciences, that if we once go wrong, errors become manifest as soon as we go a step forward, so that we have no fear of building complacently on a bad foundation, for the whole will go to the ground as soon as we make the first practical application.[21]

One of the points he later made most strongly in his Inaugural Lecture at King's College London, and it is a key to understanding the sometimes bewildering changes in tack his own research work was wont to take. A developing theory floats in the medium of experimental fact, it does not need a solid but cumbersome foundation. The interest Maxwell developed in the philosophy, history and above all the psychology of science were permanent legacies of the Apostles. He was later to write in a book review:

> As regards the most interesting of all subjects, the history of the development of scientific ideas . . .[22]

and in his copious work as a referee, time and again he asks for papers to be printed in their pristine state, so future readers could study the genesis of the work.

Another important Apostolic influence on his life was Frederick Denison Maurice, the founder of the Christian Socialist movement. In general Cambridge was religiously rather apathetic at the time, particularly by comparison with Oxford, then being torn apart by the Oxford movement. There, the 39 Articles of the Church of England had to be sworn before admission was gained to the University. At Cambridge, by contrast, things had relaxed somewhat, and the Articles were necessary only on taking a degree. One could become a Wrangler, but then fail to become a B.A. When it eventually came to pass in the eighteen sixties that a Jew became Senior Wrangler, and therefore had to take his degree alone and ahead of the rest of the pack, the University quietly passed a note allowing him to take his degree without oaths.

J.J. Sylvester, one of the great Victorian mathematicians, was also Jewish, but was only Second Wrangler in his year, (1838), despite going to St John's (George Green was fourth wrangler that year) and this problem was avoided. He could not take the Smith's Prize, nor his B.A., but he was appointed Professor of Physics at University College London, the only non-theological institution in the country. In 1841 he went to the University of Virginia as Professor of Mathematics, but after four years there he made fun of the accent of one of his students, a young man with a Southern sense of honour. The young man demanded an apology, which Sylvester refused to give. Then the man and his brother waylaid Sylvester and attacked him with a club. Sylvester had bought himself a sword-stick which he unsheathed and ran his assailant through. The young man is reported to have collapsed into the arms of his brother saying "I am killed". Sylvester fled the country, whereupon the young man recovered. On his return to Britain, Sylvester had to find a job—and he joined a firm of actuaries before becoming a barrister where he was able to spend his time talking mathematics to his friend Cayley.[23]

By the time Rayleigh came to Cambridge, the religious vows had become a bit of a farce. When he was elected a Fellow at Trinity, he was required to sign a book, which appeared merely to contain a list of all the previous Fellows. When all had signed someone asked the Vice-Chancellor whether any particular significance attached to these signatures. He replied, in some confusion, that he should have explained to them that what they had signed was the declaration that they were members of the Church of England.[24]

A decade earlier, in Maxwell's time, things were not so lackadaisical. F.D. Maurice was sacked from King's College London for heresy, and inspired many of the brightest undergraduates of the day with his ideals. He believed that the Industrial Age was de-humanising and de-Christianising the working class. Redemption was to be found in cooperatives, where the workers had a say in their labour, and through education. Working Men's Colleges were set up to give the labourer the opportunity to better himself, and the movement agitated for the early closing of shops to give the workers the free time to attend. Maxwell, a landlord and paternalistic and conservative in outlook, was inspired by the movement and until 1866 gave up at least one evening a week to teach classes in those colleges. Indeed, he seems to have preferred teaching them than to teaching pampered undergraduates:

> If I am to be up this term, I intend to addict myself rather to the working men who are getting up classes, than to pups [pupils] who are in the main a vexation.[25]

he wrote to his father.

Another interesting insight into Maxwell's views is gained from a letter to his father in 1855:

> We had a discussion and an essay by Pomeroy last Saturday about the

position of the British nation in India. . . . We seem to be in the position of having undertaken the management of India at the most critical period, when all the old institutions and religions must break up, and yet it is by no means plain how new civilisation and self-government among people so different from us is to be introduced. One thing is clear, that if we neglect them, or turn them adrift again, or simply make money of them, then we must look to Spain and the Americans for our examples of wicked management and consequent ruin.[26]

Maxwell had a strong sense of duty. It was his duty to teach at these classes, just as it had been a duty to suggest suitable books for the labourers' children to read at Glenlair. It was a religious duty for him to try and understand Nature:

> I believe, with the Westminster Divines and their predecessors *ad infinitum* that "Man's chief end is to glorify God and to enjoy him for ever." That for this end to every man has been given a progressively increasing power of communication with other creatures. . . . That happiness is indissolubly connected with the full exercise of these powers in their intended direction. . . .[27]

But as ever with Maxwell there is a sting in the tail. 'That Happiness and Misery must inevitably increase with increasing Power and Knowledge.' The question Maxwell now posed himself was whether his duty lay in Cambridge. After his examination success, he had firmly established a reputation. In Cambridge he had come to know Whewell and Stokes. William Whewell, the Master of Trinity, was the son of a Lancashire carpenter and had been forced to struggle for his early education. He had gone to Cambridge on a scholarship, been a high wrangler, and had then been successively Professor of Mineralogy and Moral Philosophy. He wrote pioneering books on the history and philosophy of science, maths text books as well as encyclopaedia articles on subjects as diverse as architecture, political economy, and, as Cambridge legend has it, Chinese musical instruments. When Faraday was trying to break away from the old electrical terminology, tied to pre-existing models of the mechanism of electricity, it was to Whewell that he turned for his neologisms:

> I am in a trouble which when it occurs at Cambridge is I understand referred by everybody in the University to you for removal and I am encouraged by the remembrance of your kindness and on Mr. Willis' suggestion to apply to you also. But I should tell you how I stand in the matter.
> I wanted some new names to express my facts in Electrical Science without involving more theory than I could help. . . .[28]

It was Whewell, too, to whom Forbes applied for help when he wanted to get modern maths introduced to Edinburgh, asking him to write a simpler version of his mechanics textbook, suitable for the low state of mathematical knowledge among Scottish students (see p.49).

George Gabriel Stokes was the Lucasian Professor of Mathematics (Newton's chair), President of the Royal Society and a Member of Parliament. He was a pivotal figure in Victorian science, both through his research, in the Cambridge applied maths tradition (but he also did experiments in his private laboratory in his rooms) and also through his committee and editorial work. Both Thomson and Maxwell bounced their new ideas off him, to see the response, which was always sensible, judicious and well-informed.

With those contacts, and with Forbes and Thomson in Scotland (Maxwell had met the latter at a British Association meeting, and started to correspond with him), Maxwell had become part of the scientific establishment. In October 1855 he became a fellow of Trinity, and the way was open for him to pursue a career in Cambridge: congenial company, good scientists with whom to discuss ideas, walking, swimming and sculling for recreation.

His father's health was causing him anxiety, however. Despite their physical separation, they had remained close, his father pestering James with practical advice, the importance of knowing the right people, the possible expense of getting working models of James' inventions made up, what to do and see on his holidays (see page 21), and even when Maxwell was studying hydro-dynamics, a research suggestion:

> Have you put a burn in fit condition to flow evenly, and not beat on its banks
> from side to side? That would be the useful practical application.[29]

Amusingly, this is a problem which later attracted the attention of no less than Albert Einstein. Now it was definitely the father who looked up to, with considerable pride, the son. In the winter of 1855/6 his father's health began to fail, and Maxwell went to Edinburgh to nurse him. He seemed to make a good recovery.

Then Forbes wrote to say that the Professor of Natural Philosophy at Marischal College, Aberdeen, had died. Was Maxwell interested in the job? This was not a ridiculous idea, most professors in those days got their post young, Stokes at twenty nine, Tait at Belfast when he was twenty three, William Rowan Hamilton got the professorship of astronomy in Dublin when he was twenty two, and still an undergraduate, and William Thomson was the same age when he went to Glasgow, where he proceeded to sit for nigh on sixty years.

Maxwell decided to apply for the Aberdeen post. As well as being closer to his father, he felt that Cambridge, staffed by its celibate bachelor fellows was an ivory tower, and that Aberdeen would be closer to real life:

> The transition state from a man into a Don must come at last, and it must be
> painful, like gradual outrooting of nerves. When it is done there is no more pain,
> but occasional reminders from some suckers, tap-roots, or other remnants of the
> old nerves, just to show what was there and what might have been.[30]

CHAPTER 7

In democratic tradition, the Scottish universities' year ran from November to April, to allow students to go home and help during the farming season. This left the professors free to get on with their research, travel or hunting. For Maxwell this was an extra attraction, since it would let him live at Glenlair in the busiest months of the year, and shoulder his share of the laird's responsibilities for his father.

So Maxwell set about the task of getting eminent men to write fulsome letters of praise, and moreover, to write them soon, a task he carried out with considerable effect, though not without a wry delight in the whole process:

> ... if you believe the Testimonials you would think the Government had in their hands the triumph or downfall of education generally, according as they elected one or not.[1]

The humour of the situation was heightened by one of the other candidates for the post applying to Maxwell himself for a testimonial, Another applicant for the job was Maxwell's schoolfriend, P.G. Tait, already Professor of Mathematics in Belfast, but who had now decided that he wished to concentrate on physics.

It was Maxwell who was appointed, but his father, who would have enjoyed the triumph more than James himself, unfortunately did not live to see it. Less than a month before the announcement, he died suddenly, on 3 April 1856 after a morning's gardening at Glenlair. It happened during the Easter vacation, and his son was by his side.

It was a great blow to Maxwell, to which he reacted in typical stoic fashion by burying himself in work: he returned to Cambridge as soon as possible to resume his lecturing duties, and to 'overtake' some 'stiff work' he had waiting for him. At Glenlair he determined that everything should run on unchanged:

> I shall have the mark of my father's work on everything I see. Much of them is still his, and I must be in some degree his steward to take care of them. I trust that the knowledge of his plans may be a guide to me, and never a constraint.[2]

Maxwell's attachment to the estate and to family tradition and continuity was strong. He saw a sort of immortality in this continuity, in the impressions and memories left behind in the minds of family and friends.

This notion was strengthened soon afterwards by the death from exhaustion of one of closest Cambridge friends, Pomeroy, during the Indian mutiny, after he had joined the Indian Civil Service. In Cambridge Maxwell had helped nurse Pomeroy during an illness and had predicted a great career for him in India. Now he wrote:

> . . . it is in personal union with my friends that I hope to escape the despair which belongs to the contemplation of the outward aspect of things with human eyes. Either be a machine and see nothing but "phenomena," or else try to be a man, feeling your life interwoven, as it is, with many others, and strengthened by them whether in life or in death.[3]

This advanced, mystic, almost pantheistic Christianity remained a permanent feature of his religious belief to the very end. His Cambridge colleague, Professor Hort, who visited him on his death bed, recalled that he often expressed the idea that

> . . . the relation of parts to wholes pervades the invisible no less than the visible world, and that beneath the individuality which accompanies our personal life there lies hidden a deeper community of being as well as of feeling and action.[4]

For him, science was an act of worship:

> What is done by what is called myself is, I feel, done by something greater than myself in me.[5]

He said to Hort on his death bed.

Science was a personal matter between Maxwell and his creator. He was once invited to join the Victoria Institute, an organisation devoted to finding common ground between religion and science. Identical though this was to Maxwell's own search, he refused to join:

> I think that the results which each man arrives at in his attempts to harmonise his science with his Christianity ought not to be regarded as having any significance except to the man himself, and to him only for a time, and should not receive the stamp of a society.[6]

It had been his father's wish that he should go to Aberdeen, and he accepted the job despite the increase in his duties at Glenlair. He resigned his Cambridge fellowship, and returned to Scotland. The somewhat unsure and uncertain adolescent who had left Edinburgh six years earlier had matured to a confident and quietly determined man. He was a full professor now, with all the responsibilities that that entailed, with some good research already behind him, and was poised on the brink of some great work. At Cambridge he had published his first paper on Electromagnetism, and this work continued until the eighteen-seventies. He took up gas theory at Aberdeen, and continued to do crucial research in this field until his death in 1879.

There is a logical sequence and flow to these strands of his work, spread over periods of twenty years, that operate quite outside the separate but entwined strand of his life. They will therefore be dealt with separately, the main themes in his research will be treated in their entirety in later chapters, and for the moment it is his everyday life that will be focussed upon. This division of the man from his work, is, of course, spurious (even though it may be unavoidable) particularly with a man who was as dedicated to his work as Maxwell was.

A long-sustained research effort also runs against the standard idea of genius. Popular legend likes to concentrate on the moment of inspiration in science: Newton, Archimedes, Fleming . . . even Einstein was prepared to acquiesce. He once humourously told journalists that he had the idea for the General Theory of Relativity from seeing a workman fall off the roof of a building. The man was, luckily, unhurt, and when asked afterwards about the experience, said that while he was falling he had not felt a thing—a statement equivalent to one of the fundamental postulates of General Relativity. What Einstein did not say was that he had put forward that postulate over ten years earlier. Maxwell did not play that game.

In his private correspondence, though, Maxwell did leave the occasional tantalising hint of his working methods. His first, Cambridge, paper on Electromagnetism was intended merely as an opening salvo. During his Aberdeen period, he kept intending to return to the subject, there was even one subterranean rumble when he wrote that he felt, 'the electrical state come on again,[7] but Aberdeen was a period of electrical 'fermentation'.

He was kept busy enough in other ways. In Cambridge he had given a few specialist, advanced courses of lectures. In Aberdeen he was responsible for the entire Natural Philosophy course, given to students whose knowledge was more rudimentary, and in addition he undertook some Working Men's Lectures. 'I have 15 hours a week, which is a great deal of talking straight forward', he wrote. Then there were weekly examinations, both oral and written, involving both the setting and marking of questions—Scottish universities were very exam-conscious.

> So I have got into regular ways, and have every man *viva voce'd* once a week, and the whole class examined in writing on Tuesdays, and roundly and sharply abused on Wednesday morning.[8]

There were also practical demonstrations to prepare. For the last of his lectures on fluids, the class went outside and Maxwell 'finished off with a splendid fountain in the sunlight. We were not very wet'.

Professor R. V. Jones, the present incumbent of Maxwell's chair at Aberdeen, found that one of Maxwell's demonstrations was still remembered in Aberdeen. He had mounted one of students on a stand and 'pumpit him fu' o' electricity, so that his hair stood oot on end'.

The practical demonstrations worked well:

> . . . the class could explain and work out the results better than I expected. Next year I intend to mix experimental physics with mechanics, devoting Tuesday and Thursday (what would Stokes say?) to the science of experimenting accurately.[9]

The idea of students doing their own experimental work was a revolutionary one; until then, if the students were lucky, the lecturer would stage-manage a few carefully chosen demonstrations in front of them. It was an important contribution to the teaching of physics to get the students to do their own experiments, and one which Maxwell later pursued at King's College London, and brought to fruition when he finally returned to Cambridge as the Cavendish Professor of *Experimental* Physics.

During term time, Maxwell had a heavy work load. Much of his leisure was devoted to research on the structure of Saturn's rings. This topic had been set as the subject for the Adams' prize essay offered by St. John's College Cambridge in honour of one of the most celebrated of Johnian Wranglers, John Couch Adams, who by dint of mind-numbing calculations had worked out that the solar system must contain a hitherto undiscovered extra planet, whose gravitational effect was disturbing the orbits of the other planets. He had even calculated where it should be in the sky, and there the astronomers had found Neptune.

The prize topic was a suitable one, for it involved reams of heavy calculations, which Maxwell had to wade through:

> I find I get fonder of metaphysics and less of calculation continually. [10]

He persevered, and won the prize, proving that Saturn's rings could not be solid or fluid, but instead had to be made of particles. The effort was worthwhile, for the paper established him not merely as a promising young scientist, but as a fully-fledged leader of the scientific community. 'One of the most remarkable applications of Mathematics to physics that I have ever seen', said the Astronomer Royal, George Biddell Airy (the capitals are his too). [11]

Working hard for the six months of the academic year, and then disappearing to Glenlair for the summer gave his life an insubstantial and very lonely quality at Aberdeen. He never bought a house there, but lived the whole time in digs. For exercise he had always preferred solitary sports:

> I had a glorious solitary walk to-day in Kincardineshire by the coast—black cliffs and white breakers. I took my second dip this season. I have found a splendid place, sheltered and safe, with gymnastics on a pole afterwards. [12]

Anyone who has ever tentatively dabbled a toe in the sea off Scotland at the height of summer will find it quite remarkable that that letter was written on 27 February 1857.

Added to those factors, Maxwell had to uphold his new-found professorial *gravitas*, as a Professor of Natural Philosophy. He found no one in Aberdeen with whom he could relax and let his imagination and humour run riot. His normal attitude was that:

> Tho' all things are full of jokes, that does not hinder them from being quite full, or even more so, of more solemn matters [13]

but he could not trust any of his Aberdeen acquaintances to appreciate this. He wrote to Campbell that he was thankful he had not tried to 'mystify' anyone there.

He saved his practical joking for Cambridge. When he went there in 1857 to collect his M.A. he took with him an improved version of the top he had used in his experiments on colour with Forbes. He demonstrated the new model one evening to his friends, and it was still rotating when they left. In the morning he saw them through his window returning to take him in to breakfast, and so set it spinning again, and got back into bed. They began to suspect that he had discovered perpetual motion. But in Aberdeen, things were decidely different:

> Gaiety is just beginning here again. Society is pretty steady in this latitude,— plenty of diversity, but little of great merit or demerit,—honest on the whole, and not vulgar . . . No jokes of any kind are understood here. I have not made one for two months, and if I feel one coming I shall bite my tongue. [14]

One of the least humourous of the people he knew was the Principal of Marischal College, the Very Reverend Daniel Dewar. Every Thursday afternoon there was a compulsory lecture known as the 'Myrtle Lecture' a sort of weekday religious service for the students and staff, which Maxwell had to attend in his official capacity. The Principal would drone on interminably in a 'deep raucous voice' on the 'instability of human greatness' or on a similar worthy theme, quite unaware that his audience were staging a minor riot. Maxwell composed this little ditty for Campbell.

> Know ye the Hall where the birch and the myrtle
> Are emblems of things half profane, half divine,
> Where the hiss of the serpent, the coo of the turtle,
> Are counted cheap fun at a sixpenny fine?
> Know ye the Hall of the pulpit and form,
> With its air ever mouldy, its stove never warm;
> Where the chill blasts of Eurus, oppressed with the stench,
> Wax faint at the window, and strong at the bench;
> Where Tertian and Semi are hot in dispute,
> And the voice of the Magistrand never is mute;[15]

One can be quite sure that this was not for Aberdonian consumption, for though Dewar might have been humourless, he was also a kindly man, and frequently invited the solitary Maxwell to his house. Maxwell fell in love with his daughter, Katherine, and they were married in June 1858.

In the best traditions of Victorian biography, Campbell says nothing about the marriage, except that the couple were wonderfully happy, and quotes some letters of quite stultifying piety to prove it. Katherine Mary Dewar was seven years older than Maxwell, she suffered from bad health throughout their marriage, and Maxwell had to be her constant nursemaid, even when he himself was terminally ill. This fact alone could explain why to Maxwell's colleagues, she seemed jealous and possessive, and resented the time he spent with others. In his biography of his father, *the* Lord Rayleigh, whose family mansion was at Terling, Robert Strutt gives this story:

> I cannot find evidence that Maxwell ever came to stay at Terling. In view of its easy accessibility from Cambridge, this seems strange. Possibly Mrs. Maxwell's unsatisfactory state of physical and mental health had something to do with it. Schuster told me (1924 or 1925) that Mrs. Maxwell had a curious jealousy of his friends. Thus Maxwell told Schuster that he would always be happy to see him at the Laboratory, but would prefer that he not call at his house. H. Sidgwick told me that he had taken walks with Maxwell. He found that Maxwell's train of thought ran on very fast. The connection of his remarks with the previous conversation could be perceived, but not without an effort.[16]

This is comparatively mild compared to some of the stories that have circulated about Mrs. Maxwell. Another was that she was of a jealous disposition and if ever her husband seemed to be enjoying a social gathering too much she removed him at once. It became so blatant that once at a dinner party at the Kelvin's when Professor Crum Brown was laughing and talking at the very top of his form, Lord Kelvin said to Mrs. Crum Brown "Don't you see he is enjoying himself? Take him away" supposedly in imitation of Mrs. Maxwell at her worst.[17]

It is difficult to know now how far to trust such stories. This latter was spread by Mrs. Tait who was notorious for her dislike of Mrs Maxwell. Another story that is almost certainly wrong is that she is supposed to have fancied the role of a laird's wife, and wanted Maxwell to give up academic life for one of hunting, fishing and shooting. Mrs. Maxwell was brought up to an academic background, and there is evidence that she was *petrified* of her duties as laird's wife, at least initially. Professor C. W. F. Everitt has discovered that Maxwell had earlier fallen in love with his cousin, Elizabeth Cay, a beautiful and intelligent girl, and they planned to marry. But family pressure had convinced them of the perils of consanguinity and they had terminated their affair. Mrs. Maxwell may have come to hear of this, which would not have improved her peace of mind.

Mrs. Maxwell may well have resented the time Maxwell spent on his physics rather than with her, but she seems to have been a willing assistant in his experiments when asked. Indeed she was rather useful. In his work on colour vision, Maxwell found that practically everybody saw a particular optical effect, the Maxwell spot, under appropriate conditions of illumination. His wife was almost unique in that she did not—a fact which Maxwell correctly attributed to the lack of yellow pigment in her retina. (Katherine makes a scientific debut in Maxwell's life long before he mentioned her in his letters. In his research on vision the missing yellow pigment is attributed to the Kafkaesque Subject K).

Later on, when Maxwell was doing important experiments on the viscosity of air at different temperatures, Katherine again assisted:

> For some days a large fire was kept up in the room, though it was in the midst of very hot weather. Kettles were kept on the fire. and large quantities of steam allowed to flow into the room. Mrs. Maxwell acted as stoker, which was very exhausting work when maintained for several consecutive hours. After this the room was kept cool, for subsequent experiments, by the employment of a considerable amount of ice.[18]

Campbell wrote that, but he was no scientist. In the published record of this experiment, Maxwell shows the apparatus surounded by a steam jacket, so that steam could be passed in without boiling the experimenter *en route*. Some steam might have escaped, it is true, but for the low temperature experiments it is hardly likely that the whole room would have been cooled.

Mrs. Maxwell also went through considerable personal discomfort to nurse Maxwell during his illnesses. In 1860, Maxwell caught smallpox while buying a pony they called Charlie, for his wife at the Rood fair in Galloway. She sat beside him and nursed him continuously through the course of the disease, no one else being allowed into the room. She did the same thing during an attack of erysipelas he got from a scratch on the head while out riding.

Maxwell does not seem to have been lucky near horses, nor was he particularly lucky at Aberdeen. There were then two separate universities there, Marischal College and King's College. It was becoming obvious to all that this was not a viable arrangement and some sort of merger was necessary. The choice was between union and fusion: union where both sets of professors would stay on, fusion where only one professor per subject would be retained. Public opinion felt that with two professors competing for students (who had to *pay* to attend lectures) the professors would be biddable and would teach what was popular. Maxwell was firmly on the side of academic freedom, and proclaimed himself a fusionist. This was a topic close to his heart. He wrote to Thomson at around this time about a mooted 'Board of Examiners for Scotch Colleges', 'to ensure a uniform standard of excellence'. But from his Tripos experience at Cambridge, he felt its effect would only be to crush out all individuality and interest from the lectures.

Fusion won the day, but with unfortunate consequences for Maxwell. His counterpart at King's College was David Thomson, a nephew of Michael Faraday, and who had taught William Thomson for a time at Glasgow. He too was a fusionist, but was in addition a much better politician than Maxwell (he was known locally as 'Crafty Thomson'). The general rule on fusion was that the younger of the two professors would be kept on, while the senior retired, based on the sound economic principle that since a life pension was to be paid to whichever professor retired, the senior professors were likely to die sooner and therefore be cheaper. This rule was applied in every case but one: Thomson kept his chair and it was Maxwell who was retired in 1860.

In the same year, James Forbes, who had been ill for a while, resigned from his chair at Edinburgh—to take his retirement as the Principal of the University of St. Andrews. Maxwell now applied for the vacant chair, but this time the Aberdeen results were reversed and P. G. Tait defeated him for the chair. In his years at Aberdeen Maxwell had published his theory of Saturn's rings, and done brilliant work on gas theory. The rumour circulated, however, that the choice had gone against him because of his poor performance as a teacher, as a percipient commentator in the *Edinburgh Courant* made plain:

> . . . in Professor Maxwell the curators would have had the opportunity of associating with the University one who is already acknowledged to be one of the remarkable men known to the scientific world. His original investigations on the nature of colours, on the mechanical condition of stability of Saturn's rings, and many similar subjects, have well established his name among scientific men; while the almost intuitive accuracy of his ideas would give his connection with a chair of natural philosophy one advantage, namely that of a sure and valuable guide to those who came with partial knowledge requiring direction and precision. But there is another power which is desirable in a professor of a University with a system like ours, and that is, the power of oral exposition proceeding upon the supposition of a previous imperfect knowledge, or even total ignorance, of the study on the part of the pupils. We little doubt that it was the deficiency of this power in Professor Maxwell principally which made the curators prefer Mr. Tait . . . [19]

It would also appear that Edinburgh's pride was at stake: they were not going to take anyone else's rejects.[20]

The question of Maxwell's lecturing ability is a complicated one. Surely someone who could write so vividly, and had such a clear and firm grasp of physics should be able to give lectures. The texts of his popular articles are superb, as indeed are the texts of his carefully prepared official lectures. But these are special efforts for which he restricted himself to a script.

Ordinary lecturing was somewhat different. The way he mixed lectures with tests and exams to keep the students alert shows that he was not unaware of some of the problems of lecturing, as he wrote to Thomson from Aberdeen:

> At present the hour from 9 to 10 is supposed to be oral exam and 11 to 12 lecture but I find it best to do both at both hours and examine without warning, for pure examination is tiresome for those who are not examined and pure lecturing encourages passivity in passive men, not to say talking and notewriting among the oblique minded.[21]

Maxwell was able to produce testimonials attesting to the success of his methods, both from colleagues at Aberdeen:

> As a teacher I believe that Professor Maxwell has ever been successful, since entering on the duties of his office in the Marischal College and University of Aberdeen.
>
> It is not, however, merely as a Lecturer on Natural Philosophy that he has given satisfaction to the students, but as combining with his daily lecture a tutorial system of teaching, carried out by weekly examinations, so as to test the diligence and progressive attainments of every student in his class; dictating questions which are answered in writing (extempore) on each of the subjects to which reference has been made in the course of the preceding week. By this latter method—in which Professor Maxwell especially excels—he has, I am satisfied, succeeded in communicating to every intelligent mind among his students a clear knowledge of the subject, and inspiring not a few of them with a decided taste for the cultivation of physical science in after life. With the maintenance of authority in his class Professor Maxwell unites an unaffected and kindly intercourse with his pupils, which contributes also, in no small degree, to his success as an instructor.[22]

and from the contented fathers of some of his prize pupils:

> Having heard of your intention to come forward as a candidate for a Chair in our metropolitan University, I have much pleasure in testifying to the benefit derived from your instructions by my son, who carried the first prize in the Natural Philosophy Class in Marischal College last winter.
>
> To these instructions he was greatly indebted. The accounts given by him of your clear and interesting illustrations of the various subjects discussed, were in keeping with what I had myself observed of your explanations and illustrations of similar matters in conversation. I know that you have gained and secured the affections of your classes; while in not a few of your pupils you have awakened what promises to be more than a passing interest in that branch of science in which you were required to direct their studies.[23]

No one, however, ever submits *unfavourable* testimonials, and perhaps Maxwell's teaching practice was not up to his theory. He was chronically incapable of gearing his mind down to meet the pedestrian needs of the average student. The problem lay in the rapidity of his own thought (see p.28). He had himself never experienced any difficulty in keeping up with his lecturers at Cambridge, as a contemporary described:

> I have sometimes watched his countenance in the lecture-room. It was quite a study—there was the look of a bright intellect, an entire concentration on the

subject, and sometimes a slight smile on the fine expressive mouth, as some point came out clearly before him, or some amusing fancy flitted across his imagination. He used to profess a dislike to reproducing speculations from books, or hearing opinions quoted taken bodily from books.[24]

Although he realised his own speed of thought could be a difficulty for others (he advised a friend who was having problems conveying his message in his sermons to a country congregation "Why don't you give it them thinner?") he was unable to do anything about it himself, as Campbell knew:

> A hindrance lay in the very richness of his imagination, and the swiftness of his wit. The ideas with which his mind was teeming were perpetually intersecting, and their interferences, like those of the waves of light, made "dark bands" in the place of colour, to the unassisted eye. Illustrations of *ignotum per ignotius*, or of the abstruse by some unobserved property of the familiar, were multiplied with dazzling rapidity. Then the spirit of indirectness and paradox, though he was aware of its dangers, would often take possession of him against his will, and either from shyness, or momentary excitement, or the despair of making himself understood, would land him in "chaotic statements", breaking off with some quirk of ironical humour.[25]

It was a problem he never overcame, as Campbell found when talking to a Cambridge colleague of Maxwell in the 1870s:

> I have been told by Mr. Huddlestone, a late Fellow of King's College, Cambridge, that when consulted about a lightning conductor for King's College Chapel (a building which he greatly loved) Maxwell called and made a verbal explanation which was unintelligible, but in going away he fortunately left a written statement, and this was perfectly clear.[26]

On the other hand, for the gifted and/or the interested student his lectures must have been wonderful. The very features which made him a poor lecturer to the average student would stimulate the very good: the sight of a brilliant mind churning out ideas and associations like sparks from a grindstone must have been inspiring. It certainly was to Sir David Gill, one of his students at Aberdeen, later to become the director of the observatory at the Cape of Good Hope:

> After the lectures, however, Clerk Maxwell used to remain in the lectureroom for hours, with some three or four of us who desired to ask questions or discuss any points suggested by himself or by ourselves, and would show us models of apparatus he had contrived and was experimenting with at the time, such as his precessional top, colour box, etc. These were hours of purest delight to me.
>
> Maxwell's lectures were, as a rule, most carefully arranged and written out—practically in a form fit for printing—and he would begin reading his manuscript, but at the end of five minutes or so he would stop, remarking, "Perhaps I might explain this," and then he would run off after some idea which had just flashed upon his mind, thinking aloud as he covered the blackboard with figures and symbols, and generally outrunning the comprehension of the best of us. Then he would return to his manuscript, but by this time the lecture hour was nearly over and the remainder of the subject was dropped or carried over to another day. Perhaps there were a few experimental illustrations—and they very often failed—and to many it seemed that Clerk Maxwell was not a very good Professor. But to those who could catch a few of the sparks that flashed as he

thought aloud at the blackboard in lecture, or when he twinkled with wit and suggestion in after lecture conversation, Maxwell was supreme as an inspiration.[27]

Edinburgh's loss was London's gain. The professorship of Natural Philosophy at King's College London was also vacant, and Maxwell was appointed to the job in the summer of 1860.

CHAPTER 8

Maxwell's first duty at King's was to deliver an inaugural lecture, a formal affair attended by the Principal of the college. Maxwell gave his views on education and metaphysics, and a summary of the current state of physics. He then rounded off his review:

> Last of all we have the Electrical and Magnetic sciences, which treat of certain phenomena of attraction, heat, light and chemical action, depending on conditions of matter, of which we have as yet only a partial and provisional knowledge. An immense mass of facts has been collected and these have been reduced to order, and expressed as the results of a number of experimental laws, but the form under which these laws are ultimately to appear as deduced from central principles, is as yet uncertain. The present generation has no right to complain of the great discoveries already made, as if they left no room for further enterprise. They have only given Science a wider boundary, and we have not only to reduce to order the regions already conquered, but to keep up constant operations on the frontier, on a continually increasing scale.[1]

The extraordinary thing about this, is that while he was at King's, Maxwell did take the laws of electricity and magnetism, and re-express them in the form by which we know them today: Maxwell's equations. In this new synthesis he not only unified two age-old branches of physics (electricity and magnetism) into a conceptual whole, but then showed that his equations predicted an electromagnetic wave motion identical to light. Optics, a third distinct branch of physics since the beginnings of science, was now integrated as a sub-division of the grand electromagnetic unification.

Maxwell's stay at King's was the most fertile period in his life. In addition to hiw work on electromagnetism, he did both theoretical and experimental work on the viscosity of gases, continued his researches into colour vision and took the world's first ever colour photograph. Again, the work was carried out while he carried a heavy lecturing load: first year students on Tuesdays, Wednesdays and Thursdays at 11.30, second and third year students on the same days at 10.15, and a separate group of evening students on the Wednesdays. Of course, these lectures also required preparing and examining.

The syllabus Maxwell devised shows how he attempted to put into practice his views on education. In his inaugural lecture he attacked the Tripos-type learning.

> In this class, I hope you will learn not merely results, or formulae applicable to

cases that may possibly occur in our practice afterwards, but the principles upon which those formulae depend, and without which the formulae are mere mental rubbish.

I know the tendency of the human mind to do anything rather than think. But mental labour is not thought, and those who have with great labour acquired the habit of application, often find it much easier to get up a formula than to master a principle.[2]

He was not aiming at instant mathematical facility but at a true understanding of Nature:

The intermediate portion of mathematical science, which consists of calculation and transformation of symbolical expressions, is most essential to physical science, but it is in reality *pure mathematics*. Everything connected with the original question may be dismissed from the mind during these operations, and the mathematician to whom they are referred may be doubtful whether his results are to be applied to solid geometry, to hydrostatics or to electricity. But as we are engaged in the study of Natural Philosophy we shall endeavour to put our calculations into such a form that every step may be capable of some physical interpretation, and thus we shall exercise powers far more useful than those of mere calculation—the application of principles, and the interpretation of results.[3]

That point of view was uncannily close to Sir William Hamilton's (p.49). In the syllabus that Maxwell prepared, two things stand out. The first is his emphasis on the fundamental principles of physics, and the second is the way that the latest discoveries (often his own) were introduced into the course as it developed and adapted over the years. Sir John Randall commented in 1962 that, 'it still is a very good course'.[4] Maxwell also tried to develop his students' feeling for the practical side of physics by giving them experimental classes, thus continuing his Aberdeen policy; King's students paid a fee to use the college workshop, which can be explained only if they were actually doing experiments for themselves.

In the examinations that he set, Maxwell tried to make the questions interesting as well as challenging by embodying the physical principles he wished to examine in questions couched in real everyday experience: how to obtain measurements of the true temperature while using ordinary fallible, inaccurate equipment, or another on what appears to be the best way to become seasick through the swaying of a small boat. He insisted that the examination papers be properly printed rather than lithographed. The extra cost entailed was worth it to make sure that all the copies were legible and therefore no students were unfairly disadvantaged.

An added demand on Maxwell's time and conscientiousness was the work he did for the British Association Committee on electrical units. In the long history of the subject, an *ad hoc* collection of units had been defined, as and when researchers needed them, magnets were measured one way, static electricity another and current electricity a third—it took a long time before people were convinced that there was any connection between static and current electricity. Clearly the state of chaos that had produced 550 foot pounds weight per second in one horsepower, 6.25 gallons in a cubic foot or, best of all, 12 inches in a foot, 3 feet in a yard, 22 yards in a chain, 10 chains in a furlong and 8 furlongs in a mile, could not be allowed to extend to electricity. As Maxwell put it, the quite unnecessarily complicated conversion factors in going from, say, miles to yards, or cubic feet to gallons, were 'a fruitful source of error' to which, he might have added, he himself was especially prone.

The report the committee produced succeeded; for the first time a coherent view of the whole of electromagnetism was presented clearly and authoritatively (based on Maxwell's work, and though anonymous, that section was in all probability written by Maxwell himself). Their recommendations formed the basis of the electrical side of the International System of Units (the Gaussian system) adopted at the Electrical Congress held in Paris in 1881.

The first thing the committee had set itself to do was to provide a standard unit of resistance, since once established, copies of the standard could easily be replicated: coils of wire did not change their resistance much and were easily mass produced and transported.

The method used had been originally suggested by William Thomson, and the experiments were performed by Maxwell at King's. They involved spinning a coil at a constant rate in the earth's magnetic field. This induced a current in the coil and, in turn, this created an additional magnetic field. A small magnet with a damped suspension was hung in the middle of the coil, and its equilibrium position, i.e. where the effects of the earth's field and the coil's field balanced, was measured. A theoretical analysis showed that the magnitude of the earth's field was irrelevant, and the only unknown in the equation was the coil's resistance.

Because it was vital that the measurements should be as accurate as possible, all precautions to prevent error had to be taken, and the experiments became extremely long drawn out. The coils had to be spun by hand at a constant rotation rate for periods of up to twenty minutes, and as many others have done before and since, Maxwell found Murphy's law to override all other physical effects:

> 28th January 1864.
> We are going to have a spin with Balfour Stewart tomorrow. I hope we shall have no accidents, for it puts off time so when anything works wrong, and we cannot at first find out the reason, or when a string breaks, and the whole spin has to begin again. . . .[5]

The need for accuracy led Maxwell along bizarre paths: while the magnitude of the earth's field was irrelevant, its direction was not. The direction of the North Magnetic Pole wandered fractionally by the hour, and had to be allowed for in long experiments. Magnetic disturbances caused by the passage of ironclad steamers on the Thames nearby (King's College is in the Strand, and almost backs onto the Thames) visibly upset his apparatus. Then, finally, having measured the coil's resistance, came the problem of measuring the length of the wire in the coil to get the crucial value of the resistance per unit length. He had to hold the uncoiled wire straight, to get its length without straining it, and solved the problem by finding a long room (a gallery in the college museum proved ideal) and laying the wire in the grooves between the floorboards. The experiments took two years to complete and were not helped by the fact that Maxwell's collaborator appointed by the B.A. Committee was Fleeming Jenkin (also an old boy of Edinburgh Academy) who lived in Edinburgh. Daily progress reports had to be written at King's and posted North. Maxwell's workload became so great that King's appointed a lecturer in Natural Philosophy to 'lighten the labours of the Professor'.

Later on, in Cambridge, Lord Rayleigh decided to redetermine the standard Ohm, using the same apparatus as Maxwell, though with an improved method of keeping the rotation of the coil constant. When it came to analyse the data, Rayleigh found that Maxwell, true to his word, had at one critical point in calculating an electrical parameter of the coil (its self-inductance, in fact) got the coil's height and radius confused. Rayleigh's son and biographer wrote:

> There was no doubt that a confusion of this kind had arisen, and it is a striking lesson in the value of unambiguous language, at any rate outside the field of politics.[6]

The work of the committee did not stop after a standard of resistance had been defined, they went on to consider the ratio of the electromagnetic to the electrostatic units of charge. Because Maxwell had predicted that electromagnetic waves would travel with a speed equal to this ratio, it suddenly became crucial to check how closely this ratio matched the speed of light.

The Committee considered three possible avenues of attack, before Maxwell settled on another which he deemed best. Then he and Charles Hockin, a fellow of St. John's College in Cambridge, set about the epic experiment. This was to balance the electrostatic attraction of two condenser plates against the magnetic repulsion of two coils of the same size. One plate and coil was suspended on the arm of a torsion balance, with a compensating arm to cancel the effect of the earth's field. The experiment was performed in the private laboratory of John Peter Gassiot, in his house on Clapham Common. Gassiot had made a considerable fortune from the import of port wine, and this enabled him to indulge in his science as a hobby.

He equipped his private laboratory with the very best apparatus available, and did some good work—he was the Bakerian lecturer at the Royal Society in 1858. Some of his experiments had required very high voltages, and so he had had constructed huge batteries. This was exactly what Maxwell now needed and he borrowed one consisting of 2,600 cells. It is a fair reflection on the state of science at the time, that an amateur scientist and professional wine importer could afford to run the biggest and best electrical laboratory in the country. The value Maxwell and Hockin obtained in 1868 for the ratio of the two units of electrical charge was 288,000 kilometres per second, fairly inaccurate, but it confirmed earlier results showing it was very close to the speed of light.

King's College had been founded in 1828 as an Anglican answer to University College, London, which was itself a non-conformist answer to Oxbridge. From the beginning such progressive and revolutionary subjects as chemistry, botany, economics, geography and experimental philosophy (physics) were taught, as well as professional courses in engineering, law and medicine. The second quarter of the nineteenth century saw a revolution in British education. Not only were London and Durham Universities founded, but Mechanics' Institutes sprang up all over the country to attempt to give technical training to artisans (Maxwell sacrificed his evenings to teach at the Working Men's Colleges). This was partly a response to the 'Decline of British Science' movement.

The seeds of the Decline, if such it was, were sown by the very successes of Britain in the eighteenth century. Joseph Black's academically interesting discovery of latent heat was followed by Watt's practical application of the finding, and led to a vast commercial exploitation and to the Industrial Revolution. The relationship between science and technology could not be clearer. But since Britain's industry had succeeded in taking advantage of the discovery, the *laissez-faire* organisation of British industry and education at that time was obviously the right one, and this was proved by British military success abroad. The fact was, though, that in the Napoleonic War French naval technology was superior to the British: their ships were better, but their officers and men were not so stoic in the face of limb-mangling broadsides of cannon-fire. Nonetheless, Nelson did win, Britain was Top Nation, and so it was obvious that it already had the best of all possible industrial and education systems.

The conquered nations of Europe observed Britain's industrial supremacy and drew the correct conclusions. Napoleon instituted the *Grandes Ecoles*, to produce technologists and administrators. Furthermore the fact that France had (largely) lost its empire, meant that it no longer obtained cheap goods from abroad nor did it have a captive market on which to dump its produce. There was thus a tremendous incentive to develop efficient and competitive processes to manufacture goods. The soda, sugar-beet, saltpetre, steel, and bleach industries all benefitted.

In Prussia the same need for technical education was felt. William von Humboldt (brother to Alexander) was appointed Director of Education in 1809. As well as founding Berlin University, he created the *Technische Hochschule*, specifically aimed at turning out technologically-educated industrial scientists and engineers. There was also set up a system of universal education the *Realschule* and *Gymnasiums*. The Prussian School system was given over to the Pestalozzi educational system after Jena. In 1854 this system was banned for 'while much stilted talk was used both about the children and to the children, it was found that, in many cases, they were suffered to go through school without learning to read and write'[7] and more *traditional* educational methods were reintroduced.

The success of those continental developments was then noticed in Britain, the most vociferous voice being raised by Charles Babbage. He is now mostly remembered as the father of the computer, designing and partly building a mechanical calculating machine far in advance of its time, the 'analytical engine'. Babbage was also for a time the Lucasian Professor of Mathematics, and he and his cohorts revolutionised Cambridge mathematics. Until his time mathematics had been ossified, slavishly following the fluxion (or dot) notation of Newton rather than the differential (d–) notation of Leibniz and the continental mathematicians, which was more flexible, sophisticated and easier. Babbage and two like-minded rebels, John Herschel (son of William Herschel, musician, composer, deserter from the German army, amateur astronomer and discoverer of Uranus, in 1781) who had been Senior Wrangler in 1812 and George Peacock who had then been second Wrangler (Babbage did not take the exam, because, it is said, he did not want to be beaten by the other two) together decided, as they put it, to take Cambridge out of its dot-age, and convert it to the true d-ism of the continent.

In fact this proved surprisingly easy. Each year the Tripos was set by two examiners, with the assistance of two moderators who would take over the reins the next year. Because of the limited tenure of most fellows, the examiners were mainly young men, ex-top Wranglers, still in their twenties. All the young guard had to do, then, was to take control of the Tripos, set d-istic problems and rest assured that the eagle-eyed tutors would teach their students continental analysis.

The speed of response of the Tripos was its great redeeming feature. The need to find vast numbers of questions every year often led the examiners out of desperation to include problems relating to their own current interests. Stokes set as a Smith's Prize question in Maxwell's year the proof of a then unknown formula, but which has since become a standard theorem in vector analysis, known as 'Stokes' theorem'. Maxwell as examiner for the Tripos in 1869 set the candidates to prove a theorem about the dispersion of light. He made no fuss about it, nor did the examinees, so the formula has entered the textbooks under the name of its later rediscoverer, as Sellmeyer's formula. Sometimes the examiners got unexpected answers: the story is that Stokes invented the 'Stokes vector' for the analysis of polarised light as the answer to a Tripos question in the year he sat it.

His task accomplished here, Babbage next turned to the Royal Society. When originally founded, it had the practical aim of turning 'philosophy into one of the arts of life of which men may see there is daily need'. They studied town planning (Wren rebuilt London after the Great Fire), the improvement of silk manufacture, how to make iron with sea coal. But those ideals had long since been lost, and under the extended presidency of Joseph Banks the Royal Society had become another London Club, whose aristocratic members outnumbered the scientific ones. Joseph Banks had first achieved prominence as the botanist who accompanied Captain Cook on the 'Endeavour' on his first great voyage of exploration. Banks originally intended to go on the second voyage also, but demanded to take with him so much equipment, and required so much manpower and to be treated with so much pomp, that the Admiralty eventually told him to get lost. Banks then chartered a whole ship of his own and went and explored Iceland instead.

The problem Babbage faced was much more difficult than with the Tripos, since old Royal Society members did not simply fade away after seven years, but clung on grimly to their privileges and their vote. Nonetheless Babbage almost made it. When a new R.S. president was needed, he put up his friend

John Herschel for the job, while the establishment ran H.R.H. the Duke of Sussex. In a close election, the Duke of Sussex just won, because, it is said, Babbage was so confident of victory that he told some of his far-flung supporters not to bother coming to London to vote. Although this battle was lost, the war had been won, for then the Royal Society set about reforming itself.

Babbage was not to know that, however, and his next move was to help create the British Association for the Advancement of Science, or 'British Ass' to its friends. Meetings of scientists from different disciplines to discuss both specialised topics and others of a more generalised or political interest, had already begun on the continent. Babbage had attended one such, while travelling for reasons of health. The trip also enabled him to gather material to show how Europe had overtaken Britain in its interest in, and application of, science. On his return home, he wrote a book attacking the Royal Society and every other aspect of science in Britain, *Observations on the Decline of Science in Britain* (1830). It raised a storm. Fellows of the Royal Society repudiated it in the letter columns of *The Times*, thus fanning the flames. Sir David Brewster in Edinburgh lauded it in *The Quarterly Review*, and the positive outcome of Babbage's and Brewster's polemics was the foundation of the 'British Ass'.

Its effect was galvanising. The Association held an annual meeting, each year in a new venue (deliberately not in London) where the very top figures in science went and talked. It is hard today to appreciate the impact of the Association. The enthusiasm and interest generated were extraordinary, although perhaps this was not altogether surprising when, for instance, during the debate on the burning question of evolution heavyweights such as Bishop Wilberforce (a Wrangler bishop) and Thomas Huxley waded into the arena. John Richard Green, a journalist, was present at that formidable meeting:

> I haven't told you that story, have I? On Saturday morning I met Jenkins going to the Museum. We joined company, and he proposed going to Section D, the Zoology, etc., "to hear the Bishop of Oxford smash Darwin." "Smash Darwin! Smash the Pyramids." said I, in great wrath, and muttering something about "impertinence," which caused Jenkins to explain that "the Bishop was a first-class in mathematics, you know, and so has a right to treat on scientific matters," which of course silenced my cavils. Well, when Professor Draper had ceased his hour and a half of nasal Yankeeism, up rose "Sammivel," and proceeded to act the smasher; the white chokers, who were abundant, cheered lustily, a sot of "Pitch it into him" cheer, and the smasher got so uproarious as to pitch into Darwin's friends—Darwin being smashed—and especially Professor Huxley. Still the white chokers cheered, and the smasher rattled on. "He had been told that Professor Huxley had said that he didn't see that it mattered much to a man whether his grandfather was an ape or not. Let the learned Professor speak for himself" and the like. Which being ended—and let me say that such rot never fell from episcopal lips before—arose Huxley, young, cool, quiet, sarcastic, scientific in fact and in treatment, he gave his lordship such a smashing as he may meditate on with profit over his port at Cuddesdon. This was the exordium, "I asserted, and I repeat—that a man has no reason to be ashamed of having an ape for his grandfather. If there were an ancestor whom I

should feel shame in recalling, it would rather be a *man*, a man of restless and versatile intellect, who, not content with an equivocal success in his own sphere of activity, plunges into scientific questions with which he has no real acquaintance, only to obscure them by an aimless rhetoric, and distract the attention of his hearers from the real point at issue by eloquent digressions and skilled appeals to religious prejudice.[8]

Significantly the major scientific issues of the day were intelligible to the public, and the benefits wrought by science were immediately obvious: antiseptics, anaesthetics, telegraphy. The British Association popularised science in the best possible way, it brought the top researchers into contact with the public, and brought home the importance of their work.

It was a genuinely popular movement too. In August 1857 a public meeting in Aberdeen decided to invite the British Association annual meeting, but when they realised there was no suitable site, the citizens of Aberdeen raised money by public subscription for a hall to house the B.A. It was built within two years, and it was at the B.A. Aberdeen meeting of 1859 that Maxwell read his first, revolutionary paper on gas theory, in which he derived the formula for the distribution of molecular velocities in a gas. It was here also that Sir Charles Lyell in his Presidential Address to the B.A. Geological Section made an important announcement:

> . . . a work will shortly appear by Mr. Charles Darwin, the result of twenty years of observations and experiments in Zoology, Botany, and Geology, by which he has been led to the conclusion that those powers of nature which give rise to races and permanent varieties in animals and plants, are the same as those which in much longer periods produce species . . .[9]

This was the first *public* intimation of the Origin of the Species, and Evolution. The importance of the B.A. is further emphasised by the fact that it was the B.A. and not the Royal Society which set up the committee for the standardisation of electrical measurements.

But if the B.A. flourished, the new universities and institutes that had been founded to propagate science did not. The problem was more complicated than had been naively thought. Science was not really on the decline in Britain, there were more great scientists active in the middle of the nineteenth century than ever before; the problem lay in the appreciation and application of their work.

Huxley, ever ready with an apposite comment, said, 'the peculiarity of English science has been that the army has been all officers'.*

British industry failed to appreciate the change taking place in the world, the need to employ people with technical training to deal with technical jobs: the

---

* One of Huxley's most memorable statements was made about Thomson's calculations on the age of the earth:

> Mathematics may be compared to a mill of exquisite workmanship, which grinds you stuff of any degree of fineness; but, nevertheless, what you get out depends upon what you put in; and as the grandest mill in the world will not extract wheat-flour from peascod, so pages of formulae will not get a definite result out of loose data . . .[10]

Unfortunately all too many scientists have not yet learned that lesson.

need to have applied physicists and chemists to perform Watt's act of converting research into money. Because people could not get jobs even if they were qualified, they stopped bothering to acquire technical qualifications. King's College was reduced to selling off its silver spoons to remain solvent.

It was, ultimately, a question of education, but at school-level: unless there was an awareness of the importance and potential of technology inculcated in the minds of the owners of British industry while they were at school and still educable, then whatever the technical institutes did at the tertiary level was wasted. In the mid nineteenth century, Eton had twenty four classics masters, eight mathematicians, and three people to teach everything else. Lyon Playfair, professor of Chemistry at the Government School of Mines, said in 1851: 'As surely as darkness follows the setting of the Sun, so surely will England recede as a manufacturing nation, unless her industrial population becomes much more conversant with science than they are now'. Prophetic words.

The Great Exhibition of 1851 brought the facts home to those not completely dazzled by all the glassware. Though intended to enshrine the triumph of British industry, it was clear to the observant that in spite of all the destruction wrought in Europe by war and revolution, European industry was rapidly overtaking the complacent British. The classic case is the dye industry. Mauve, the first aniline dye was discovered in 1856 by W. H. Perkin at the Royal College of Chemistry. Though he made a personal fortune out of his discovery, the British organic chemical industry lost out. German manufacturers realised the potential of Perkin's discovery, and Höchst was set up in 1862, BASF in 1865. By the early 1880s, Höchst was employing fifty one chemists and BASF twenty five, all working hard to synthesise new aniline derivatives and other organic molecules to see if they would make dyes. Perkin, in evidence before a Royal Commission on Technical Education, then stated that the Germans had seventeen colour works and Britain just five, the Germans were producing two million pounds worth of dyes annually while Britain made less than half a million pounds worth. By the time of the First World War, Britain was importing all the khaki dye for its soldiers' uniforms from Germany.[11]

At the same time, Britain was preparing to lose out in the electrical industry Despite the pioneering work of Faraday, Maxwell and especially Thomson (in telegraphy, from which his patents brought him a personal fortune and a peerage) no real electrical industry developed. By the eighteen-eighties Siemens-Halske already had a small industrial physics laboratory in Britain.

It is interesting to speculate what might have happened if Justus von Liebig had come to King's College London. Liebig was the man who more than anyone was responsible for the German chemical industry. He received his training in Paris under the great French chemists of the day, and then returned to Germany to become Professor at the University of Giessen, where he developed chemical laboratory teaching, and trained almost all of the next generation of German chemists. It was his pupils who invented the dye industry in Germany. In 1845 he was tempted to become the Professor of Chemistry at King's, but, it was hastily pointed out by Bishop Blomfield, he was a Lutheran and was ineligible for appointment in the rigidly Anglican College.

Maxwell had no such problems. He was most certainly a Christian but never wholly committed to any particular division of the faith. As a schoolboy in Edinburgh he went to a Presbyterian service in the mornings with his father and to an Episcopalian service in the afternoons to please his aunt, so taking his 39 Articles and attending chapel at Cambridge did not bother him at all. He was not concerned with the superficial form of any faith, but in the mystical unity that underpins it, as he stated in his inaugural lecture at Aberdeen:

> But as Physical Science advances we see more and more that the laws of nature are not mere arbitrary and unconnected decisions of Omnipotence, but that they are essential parts of one universal system in which infinite Power serves only to reveal unsearchable Wisdom and eternal Truth. When we examine the truths of science and find that we can not only say 'This is so' but 'This must be so, for otherwise it would not be consistent with the first principles of truth'—or even when we can only say 'This ought to be so according to the analogy of nature' we should think what a great thing we are saying, when we pronounce a sentence on the laws of creation, and say they are true, or right, when judged by the principles of reason. Is it not wonderful that man's reason should be made a judge over God's works, and should measure, and weigh, and calculate, and say at last 'I understand I have discovered—It is right and true'.[12]

The sentiment was repeated almost word for word at King's. For him, then, research was almost an act of divine worship:

> Happy is the man who can recognise in the work of Today a connected portion of the work of life, and an embodiment of the work of Eternity. The foundations of his confidence are unchangeable, for he has been made a partaker of Infinity. He strenuously works out his daily enterprises, because the present is given him for a possession.[13]

Such pronouncements smack almost of pantheism, but Maxwell would have strenuously denied that. In a letter to Campbell, when an undergraduate at Cambridge, he compared the 'holy ground', the areas of dogma of different faiths:

> But there are extensive and important tracts in the territory of the Scoffer, the Pantheist, the Quietist, Formalist, Dogmatist, Sensualist, and the rest, which are openly and solemnly Tabooed, as the Polynesians say, are not to be spoken of without sacrilege.
>
> Christianity—that is, the religion of the Bible—is the only scheme or form of belief which disavows any possessions on such a tenure. Here alone all is free.[14]

He put his faith in a Christian God of creation. This statement comes from a talk about molecules to the Brittish Ass in 1873:

> . . . In the heavens we discover by their light, and by their light alone, stars so distant from each other that no material thing can ever have passed from one to another; and yet this light, which is to us the sole evidence of the existence of these distant worlds, tells us also that each of them is built up of molecules of the same kinds as those which we find on earth. A molecule of hydrogen, for example, whether in Sirius or in Arcturus, executes its vibrations in precisely the same time.

Each molecule therefore throughout the universe bears impressed upon it the stamp of a metric system as distinctly as does the metre of the Archives at Paris, or the double royal cubic of the temple of Karnac.

No theory of evolution can be formed to account for the similarity of molecules, for evolution necessarily implies continuous change, and the molecule is incapable of growth or decay, of generation or destruction.

None of the processes of Nature, since the time when Nature began, have produced the slightest difference in the properties of any molecule. We are therefore unable to ascribe either the existence of the molecules or the identity of their properties to any of the causes which we call natural.

On the other hand, the exact equality of each molecule to all others of the same kind gives it, as Sir John Herschel has well said, the essential character of a manufactured article, and precludes the idea of its being eternal and self-existent.[15]

Evolution might have something to do with the development of life on earth, but it had nothing to offer Maxwell for the solution of the fundamental problems in physics in which he was interested.

The evolution debate was tearing Victorian society apart. Many Christians thought that if Darwin was right it was the end of religion; Maxwell was serene and refused to be drawn. Evolution might affect the literal interpretation of the Bible, but not his faith. In the controversy about the age of the earth, Thomson had sound scientific reasons for his views, but nonetheless, as a regular churchgoer, he took the greatest delight in lambasting Darwin:

What then are we to think of such geological estimates as 300,000,000 years for the "denudation of the Weald?" Whether is it more probable that the physical conditions of the sun's matter differ 1,000 times more than dynamics compel us to suppose they differ from those of matter in our laboratories; or that a stormy sea, with possibly channel tides of extreme violence, should encroach on a chalk cliff 1,000 times more rapidly than Mr. Darwin's estimate of one inch per century?[16]

Darwin had quite foolishly committed himself to some wildly optimistic numbers for the duration of geological processes. He withdrew them from later editions of his work.

Maxwell did have views on evolution. Amplifying some remarks he had first made in discussion at the Cambridge Philosophical Society, he wrote to Campbell in 1874:

If atoms are finite in number, each of them being of a certain weight, then it becomes impossible that the germ from which a man is developed should contain gemmules of everything which the man is to inherit, and by which he is differentiated from other animals and men,—his father's temper, his mother's memory, his grandfather's way of blowing his nose, his arboreal ancestor's arrangement of hair on his arms, and his more remote littoral ancestor's devotion to the tide-swaying moon. Francis Galton, whose mission it seems to be to ride other men's hobbies to death, has invented the felicitous expression "structureless germs". Now if a germ, or anything else, contains in itself a power of development into some distinct thing, and if this power is purely physical, arising from the configuration and motion of parts of the germ, it is nonsense to call it structureless because the microscope does not show the structure; the germ of a rat must contain more separable parts and organs than there are drops in the sea. But if we are sure that there are not more than a few million molecules in it, each molecule being composed of component molecules, identical with those of carbon, oxygen, nitrogen, hydrogen, etc., there is no room left for the sort of structure which is required for pangenesis on purely physical principles.[17]

A very subtle criticism. Maxwell's B.A. committee collaborator Fleeming Jenkin produced an even more effective critique in a review of the 'Origin of Species' for the North British Review of 1867, in which he pointed out that any single chance mutation would be swamped out in succeeding generations, by the 'normal' individuals. Regression to the mean should be utterly dominant. One of the best descriptions of the effect of the evolution debate is in Middlemarch by George Eliot. Maxwell read it and passed this splendid parody on to Campbell:

> The Cambridge Philosophical Society have been entertained by Mr. Paley on Solar Myths, Odusseus as the Setting Sun, etc. Your Trachiniae is rather in that style, but I think Middlemarch is not a mere unconscious myth, as the Odyssey was to its author, but an elaborately conscious one, in which all the characters are intended to be astronomical or meteorological.
>
> Rosamond is evidently the Dawn. By her fascinations she draws up into her embrace the rising sun, represented as the Healer from one point of view, and the Opener of Mysteries from another; his name, Lyd Gate, being compounded of two nouns, both of which signify something which opens, as the eye-lids of the morn, and the gates of day. But as the sun-god ascends, the same clouds which emblazoned his rising, absorb all his beams, and put a stop to the early promise of enlightenment, so that he, the ascending sun, disappears from the heavens.
>
> Dorothea, on the other hand, the goddess of gifts, represents the other half of the revolution. She is at first attracted by and united to the fading glories of the days that are no more, but after passing, as the title of the last book expressly tells us, "from sunset to sunrise," we find her in union with the pioneer of the coming age, the editor . . .
>
> There is no need to refer to Nicolas Bulstrode, who evidently represents the Mithraic mystery, or to the kindly family of Garth, representing the work of nature under the rays of the sun, or to the various clergymen and doctors, who are all planets. The whole thing is, and is intended to be, a solar myth from beginning to end.[18]

Maxwell's religiosity is one of the areas where Campbell is most to be distrusted. Victorian biography always lays emphasis on this (and on happy marriages). It is amusing to compare Campbell's personal account of Maxwell as a churchgoer:

> At church he always sat preternaturally still, with one hand lightly resting on the other, not moving a muscle, however long the sermon might be. Days afterwards he would show, by some remark, that the whole service, whether good or bad, had been, as it were, photographed upon his mind.[19]

with the account in the history of Corsock Parish Church written by the man who gave those sermons, the Rev. George Sturrock:

> It may be interesting to some to know that the cushion though rather faded now, and all the books, with his own jottings on them, used in the church by

James Clerk Maxwell, still lie in the pew, just where he left them—sacred relics.[20]

Maxwell must have found time for his doodling while Campbell was not watching.

The bio-statistician, Karl Pearson, recalled, late in life, the Cambridge of his youth, and in particular what it was like to take the Tripos and Smith's Prize examinations in the eighteen seventies. Each paper for the Smith's Prize lasted a whole day, and was taken at the house of the examiner who had set it. It was broken for lunch, which the candidates ate with the examiner and his family. Pearson particularly recalled the intolerant horror with which Maxwell reacted when he had the temerity to suggest that the Biblical account of the flood might not be absolutely accurate.[21] Pearson was, however, thinking back to events which had occurred sixty years previously and I find more convincing the rather different light shed on Maxwell's religious views by Lord Rayleigh, speaking to his son:

> When I was your age I expected to attain much greater certainty than I now think can be attained. I do not think the most religious among scientific men, say Faraday or Maxwell, would pretend to certainty—perhaps Faraday would have though.[22]

While Maxwell was at King's he also got to know Faraday. They had begun to correspond while Maxwell was at Cambridge, but now Maxwell took the opportunity of meeting Faraday personally and attending his famous Friday evening lectures at the Royal Institution.

The first hint of the electromagnetic theory of light was given by Faraday at one of these seances, on 10 April 1846. Charles Wheatstone was the Professor of Experimental Philosophy at King's College, a splendid experimenter (who invented the mouth organ and the accordion, but *not* the Wheatstone bridge) but a terrible lecturer. He had obtained some very interesting results about the speed of electrical currents in wires (he thought they travelled at 288,000 miles per second), and was due to give a lecture to the R.I. As they neared the lecture theatre, Wheatstone panicked and ran off into the street, because, it is said, he had discovered at the last moment that Joseph Crabtree a notorious and inveterate heckler of lecturers was present. Faraday in the best show business tradition, strode up to the podium, and explained what he knew of Wheatstone's work. That was soon exhausted, and so he started to speculate on what would happen if his electromagnetic lines of force could vibrate:

> The view which I am so bold as to put forth considers, therefore, radiation as a high species of vibration in the lines of force which are known to connect particles and also masses together.[23]

This extemporary lecture 'Thoughts on Ray Vibrations' drew from Faraday the extraordinary speculation that radiation (i.e. light) was a transverse vibration of his lines of force.

Those ideas, which Faraday described as 'vague impressions of my mind', normally would not have been delivered. Faraday was careful not to mould his thoughts into patterns too soon by talking about them prematurely, and he was especially careful not to go on record about anything without having first checked and double-checked it. Once when a callow youth he had been badly caught out on some chemical analyses he had published hurriedly. But this lecture was delivered under exceptional circumstances.

> I do not think I should have allowed these notions to have escaped from me, had I not been led unaware, and without previous consideration, by the circumstances of the evening.[24]

Since then the Royal Institution have been most careful to prevent their lecturers doing a 'Wheatstone'. Today when a lecturer is escorted from the library to the theatre, it is reported that the Director of the R.I. is careful to interpose his body between the lecturer and his line of escape. If, however, one could guarantee a Faraday to step into the breach with brilliant extemporary insights, then all lecturers should be required to bolt!

This is one of the great stories of physics, but unfortunately it cannot be entirely true. In a recent biography of Wheatstone, Dr. Brian Bowers has checked through the Royal Institution's records, and found there was no Friday evening Discourse planned for April 10 1846—it was Good Friday in fact—and furthermore, neither Discourse subjects nor speakers were announced in advance. Faraday's notes show that he himself planned—and presumably gave—a talk on April 3 on Wheatstone's Chronoscope, and the notes end with his thoughts on ray vibrations. Bowers writes:

> Faraday must have had some advance notice that Wheatstone was not going to speak, even if Wheatstone had previously agreed to do so.[25]

Wheatstone certainly was a notoriously bad (or at least shy) public orator—though lively and friendly in private—and Faraday had had to give talks on his behalf on previous occasions.

Even the speed Wheatstone is supposed to have attributed to "The Electric Fluid"—288,000 miles per second—is unfair. This was a preliminary result which he was forced to publish by a letter from H. Fox Talbot, M.P., F.R.S. (pioneer in photography too) to the *Philosophical Magazine*. Talbot had seen some of Wheatstone's experiments and suggested in print that the apparatus could be used to measure the velocity of electricity—as if Wheatstone had not thought to do so. Wheatstone replied that this had always been his intention, and he had already got a preliminary speed. He never published a definitive value, though it seems he later hovered between 160,000 and 200,000 miles per second. Since this brackets the speed of light, and therefore ties in with Faraday's ideas, Wheatstone's reluctance to publish is extraordinary.

Despite his inability as a lecturer, Wheatstone was a great research physicist, and also, like Kelvin, made a Victorian fortune from his work on telegraphy. It is very strange that while he was the Professor of Experimental Philosophy at King's College, Maxwell was the Professor of Natural Philosophy, both had strong interests in electricity, magnetism and the speed of light, both attended the Royal Institution and were close friends of Faraday—and yet there seems to have been no interaction between them. Neither seems to have ever even mentioned the other in any of their publications or letters (such as have survived). By the time Maxwell arrived at King's, Wheatstone seems to have devoted himself entirely to research and telegraphy, and played a minimal role at King's: he gave no lectures and received no salary. Nonetheless their mutual ignorance seems odd.

Maxwell and Faraday on the other hand obviously got on well together. After one Friday evening meeting, Maxwell was caught up in a jostling melée of people all bumping and pushing into each other to get out. Faraday appreciated the similarity of the situation to Maxwell's description of the molecular motion in a gas, and shouted out "Ho, Maxwell, you cannot get out? If any man can find a way through a crowd it should be you".

Maxwell made similar use of the analogy after an appeal by W. Grylls Adams, who had been his lecturer at King's, and was then appointed Maxwell's successor (he was also the brother of John Couch Adams). Adams wanted advice on the best way to organise a scientific meeting and Maxwell told him:

> For the evolution of science by societies the main requisite is the perfect freedom of communication between each member and any one of the others who may act as a reagent.
>
> The gaseous condition is exemplified in the soiree, where the members rush about confusedly, and the only communication is during a collision, which in some instances may be prolonged by button-holing.
>
> The opposite condition, the crystalline, is shown in the lecture, where the members sit in rows, while science flows in an uninterrupted stream from a source which we take as the origin. This is radiation of science.
>
> Conduction takes place along the series of members seated round a dinner table, and fixed there for several hours, with flowers in the middle to prevent any cross currents.[26]

Adams' society is still thriving as The Physical Society.

In London the Maxwells lived at 8 Palace Gardens Terrace, a newly built house in Kensington (now number 16). The attic of the house was converted into a private laboratory where Maxwell and Mrs. Maxwell performed their experiments on the viscosity of gases, and where he terrified the neighbours by working into the night staring into a 'coffin' collecting data on colour vision. Maxwell taught and researched in the mornings mostly, the afternoons being devoted to riding in the Park—the house was, conveniently, close to Hyde Park. Despite being so central, however, one suspects that Maxwell did not much like London, he was country bred, and had a strong feeling of loyalty to Glenlair. Maxwell resigned from King's in 1865, perhaps to spend more time at Glenlair to 'stroll in the fields and fraternize with the young frogs and old water-rats'.[27]

CHAPTER 9

Maxwell had at least two other good reasons for leaving King's. One was to collate and write down his thoughts on electricity and magnetism. The outcome of this was his momentous book *A Treatise on Electricity and Magnetism*. So much of it was fundamentally new that its gestation was very protracted, it was not printed until 1873. Maxwell's textbook on Heat was also written in this period. It was published in 1870.

Maxwell also wanted to rebuild Glenlair, a long-standing obligation to his sense of feudal duty and family tradition. Maxwell later told Professor Hort that his father's plans had been a sacred trust for him:

> He wanted to build his house on a scale suited to what he thought he would require as sheriff, and had so built a small part of it when he died. We afterwards completed it, as far as possible according to his idea, but on a much smaller scale.[1]

Maxwell was of independent means but those means did not stretch far enough to fulfill his father's grandiose schemes. It was that family money which gave Maxwell the flexibility to resign. His father had said as much when he was debating whether to apply for the Aberdeen job:

> I believe there is some salary, but fees and pupils, I think, cannot be very plenty. But if the *postie* be gotten, and prove not good, it can be given up; at any rate it occupies but half the year.[2]

The heavy teaching load at King's, the work he was doing for the B.A., had bitten too much into his time, now he wanted to revert to being a scientific amateur.

Not that taking on amateur status meant slacking:

> I have now my time fully occupied with experiments and speculations of a physical kind, which I could not undertake so long as I had public duties.[3]

His strongest commitment at King's was to the evening class; despite the awkwardness of the hour, he continued lecturing to these students throughout that winter. More than just social commitment, he enjoyed teaching those who were keen to learn, and also the relative freedom he had in choosing the syllabus. There has been a legend at King's that Maxwell was asked to resign because he could not keep order in his lectures. Recent work by Professors Everitt and Domb has effectively scotched that rumour. Maxwell had valid reasons for wanting to go; the college records say that when he resigned he offered to stay on until a successor could be found; he *did* stay on to teach evening classes; and his successor, W. Grylls Adams was less able to keep order in his lectures than Maxwell. Moreover Charles Wheatstone was a totally incompetent lecturer (p.90). He was appointed professor at King's College in 1834, and remained there until his death in 1875. After his first year, when he had given a course of lectures on Sound, it seems he gave no more lectures for the rest of his tenure. Despite this, on his death the College Council decided to honour him, and resolved that "the Physical laboratory of the College should be distinguished by the name of the Wheatstone Laboratory". The rumour of Maxwell's sacking was first spread by the Rev. Canon Richard Abbay, who did not know Maxwell, but *thought* he had heard the story from Professor Clifton in 1869.

With his new-found leisure, Maxwell divided his time between Glenlair and London, with occasional trips abroad. Campbell records meeting Maxwell accidentally in Italy in 1867 'he had looked at the dome of St Peter's with an eye of sympathetic genius'. Maxwell also took the opportunity on this trip to learn Italian well, and improve his French and German. Very sensibly, the Maxwells wintered in London at a number of addresses in the Kensington area, and this is when he did his velocity of light experiment (p.81).

The summers, though, were spent at Glenlair, which he regarded as his permanent base. The house building was completed in 1867, and Maxwell undoubtedly enjoyed the role of Laird. Maxwell took a strong proprietorial interest in the choice of preacher at Corsock Church. In the summer of 1861 a new preacher had had to be found, when the incumbent moved on to a more important post. Maxwell was firm in insisting that his replacement be given a three-month trial. The Rev. Mr. Burnett started on 14 April and retired on 14 July. The next candidate for the job was George Sturrock, who came with a seal of approval from Maxwell's father-in-law, the Rev. Principal Dewar. He began on 21 July and stayed.

Maxwell's main contact with the scientific community while he was in residence at Glenlair had to be by post. He had such a weight of correspondence that he and the post office came to a mutually beneficial arrangement; they gave him a private 'pillar-box' on the road at the end of his drive. The postman did not have to tramp all the way up the muddy drive to Maxwell's front door, Maxwell did not have to walk miles down the road to post his letters. There were drafts and proofs for his books and articles to correct and recorrect in the time-honoured way for his publishers. There was an enormous load of refereeing to be done, a thankless task, which Maxwell always performed diligently; there was also the private refereeing for his friends to scan their proofs for errors. And finally there was a vast pile of ordinary correspondence to deal with.

His three main correspondents were Stokes, Thomson and P.G. Tait. Stokes seems to have been a singularly humourless man. Though he and Maxwell started exchanging letters while Maxwell was an undergraduate at Cambridge, as late as 1869 Maxwell addressed him as 'Dear Sir'. Only after he had become a Cambridge professor himself did he occasionally venture a 'My Dear Stokes', and it was not until 1875 that he attempted any levity, pointing out that 'scotch' was not the adjective for things Scottish, it was a verb, or the second syllable in butterscotch 'an English delicacy the nature of which I do not know. Perhaps the last t should be r.'[4] Like most Scots who are sticklers on this point, Maxwell himself in his less guarded moments (i.e. frequently) misused the word. Humour or no, Stokes' good sense and great knowledge made him invaluable; Thomson also used Stokes as a sounding board for his latest ideas.

Maxwell and Thomson had first met when Thomson introduced himself after Maxwell had spoken at the B.A. meeting of 1850 (see p.54). Thomson asked if Maxwell would make him some samples of the materials he had been using for his experiments on polarised light. This was recognition indeed, for Thomson was then acknowledged as the top physicist in the world! When Maxwell first became seriously interested in electricity, after taking the Tripos, it was, naturally enough, to Thomson that he wrote for advice on what to read:

> If you have in your mind any answer to the above questions, three of us here would be content to look upon an embodiment of it in writing as advice.[5]

Maxwell as acolyte. Their relationship changed to one of equality over the years, and, indeed, just before his death, it was Maxwell from the established landed family advising the *nouveau riche* Lord Kelvin about the problems of keeping ornamental peacocks in the garden:

> Mrs. Maxwell says that if you were constantly at home and combined the feeding of peacocks with your morning smoke and meditation, the birds would soon be concentrated near the centre of attraction thus provided and would not trouble the garden. Mrs. Maxwell always get the peacocks to choose the gardener and they have chosen one who has now been seven years with us.
>
> The peacocks will eat the young cabbages but the gardener tells them to go out again and they find it pleasanter to be about the house and to sit on either side of the front door.
>
> Mrs. Maxwell will not send them unless on consideration they would be acceptable.[6]

With his schoolfriend Tait, of course, there never had been any formality, and there now developed a splendid triangular correspondence between these three great Scottish physicists Thomson, Tait and Maxwell. They fired off volleys of letters to each other, filled with their latest ideas. They invited comment on their problems, invited solutions, solved problems to keep the others on their toes, and added slyly malicious gossip. Sometimes one would be having difficulty with his work in the morning, send off a letter asking for assistance, then find the answer for himself at teatime, and send off yet another letter that evening cancelling the first.

Sometimes the correspondence would be by letter, particularly if formal business had to be conducted, but often it was by ½d postcard—thus saving ½d on a letter's postage. The cramped space of the card forced a compressed style on the writers and so an intricate, punning, private language grew up. Thomson became T, Tait became T' and Maxwell became forever dp/dt after February 1871, when Tait sent him a letter starting 'Dr. J C M (= dp/dt (Tait's Thermodynamics §162))'.

In later years Maxwell used dp/dt as his *nom de plume* when writing comic verse and the odd article for the infant magazine *Nature*, published by the thrusting young Scottish publisher Alexander Macmillan. The real identity of dp/dt became one of the worst kept secrets in science. When Loschmidt and Stefan in 1872 in Vienna measured the diffusion of air and hydrogen into each other, to check Maxwell's theory, they were able to pronouce their results '*in erglanzender ubereinstimmung mit dp/dtschen Theorie*' (in brilliant agreement with dp/dt-ish theory).[7]

Here is one example of a typical card from Maxwell to Tait dealing with spherical harmonics:

> O. T'! R. U. AT 'OME? $\int\int$ Spharc^2 dS was done in the most general form in 1867. I have now bagged ξ and η from T and T' and done the numerical value of $\int\int (Y_i^{(S)})^2 dS$ in 4 lines, thus verifying T + T''s value of $\int\int (\theta_i^{(S)})$ dS. Your plan seems indept. of T and T' or of me. PUBLISH![8]

Spharc, of course, is *sph*erical *har*monic, and its plural was normally written $\Sigma\phi\alpha\rho\xi$, thus saving two letters. Similarly Stokes became $\Sigma\tau\omega\xi$ and thermodynamics $\theta\Delta$cs. Other logograms were PFian, which T explained as meaning 'Pecksniffian. Pecksniff was a great hero of Tait's in respect of his almost superhuman selfishness, cunning and hipocrisy'; Tyndall was referred to as T'' because T'' represents a tensor of the *second rank*, Macmillan as #—he was sharp, while H^2 refers to H.H., Hermann von Helmholtz, whom they respected.

Perhaps the height of their correspondence was reached in 1877 when Tait thought up, as he saw it, a crucial experiment, and urgently asked for Maxwell's views on it, by means of a 20-line poem; Maxwell's reply was as prompt as had been desired, and arrived, complete with learned references—in the form of a 48-line poem. As the historian of science, Florian Cajori, noted: 'Maxwell was fond of writing quaint verses which he brought round to his friends, "with a sly chuckle at the humour, which, though his own, no one enjoyed more than himself" '.[9]

The way in which the three physicists worked together is shown nowhere better than in the story of the vortex ring atom. Tait and Thomson, professors respectively in Edinburgh and Glasgow, were in close contact, geographically as well as in spirit, and in the early eighteen-sixties decided to collaborate to produce a book on physics based on their separate lecture courses 'blended into a harmonious whole' to be published, inevitably, by Macmillan. The book, *Natural Philosophy*, abbreviated to T and T' appeared in 1867. It is a classic of scientific exposition, but in its construction Tait noted one problem:

> A little difficulty arises at the outset, Thomson is dead against the existence of atoms; I though not a violent partisan yet find them useful in explanation—but I suppose we can mix these views well enough. . . .[10]

Thomson took the viewpoint that the existence of atoms was uncertain and it was therefore unwise to introduce the idea into a basic textbook. In the course of the eighteen-sixties, however, Maxwell published his papers on gas theory and these must have shaken even Thomson's resolve: Maxwell was able to predict many properties of gases on the atomic model.

Helmholtz had produced a classic paper on vortex motion in the late eighteen-fifties. Though the laws of hydrodynamics had been available to anyone since the time of Lagrange at the end of the eighteenth century, Helmholtz was the first person to sit down and solve them for a vortex, and show that in a liquid with no internal friction, material that starts off within the vortex always stays inside it, merely swirling around without interacting with the fluid outside. If air had no viscosity (internal friction), a smoke ring would last forever. Tait read Helmholtz' paper (in German) and was so impressed that he translated it for his own use (though he did eventually have the translation published in 1867) and he let all his friends know of the paper's revolutionary contents.

Thomson, having grown accustomed to the idea of atoms by now, and learning about vortices from Tait, put the two ideas together: if a vortex in a perfect fluid stays together as a lump forever, it makes a splendid model for the atom. He postulated that the universe was filled with a perfect fluid, just another manifestation of the aether, the philosopher's stone of Victorian physics; then the atoms were merely vortices in that aether.

Early in 1867, Tait devised a class demonstration of smoke rings. He took an open tea-chest, cut a six-inch hole in the bottom and covered the top with a taut piece of cloth, or better still rubber. By turning this on its side and tapping the cloth, vortices were shot out of the hole. They were, of course, invisible. Into the box he put a smoke machine—a dish of nitric acid and another of ammonia made a good one. Then when the box was tapped, clearly visible white smoke rings of amonium nitrate were ejected.

Tait demonstrated the device to Thomson one afternoon, perhaps when Thomson was over for a Royal Society soirée. Thomson was thunderstruck by the way that two vortices, shot out one after the other, would bounce off each other when they collided, leaving each trembling furiously: the collision of atoms was now explicable on his model, the vortex rings maintained their integrity even on collision, rather than just amalgamating.

Thomson's vortex ring paper of February 1867 thus began:

> A magnificent display of smoke-rings, which he recently had the pleasure of witnessing in Professor Tait's lecture-room, diminished by one the number of assumptions required to explain the properties of matter, on the hypothesis that all bodies are composed of vortex atoms in a perfect homogeneous liquid. Two smoke-rings were frequently seen to bound obliquely from one another, shaking violently from the effects of the shock . . . The elasticity of each smoke-ring seemed no further from perfection than might be expected in a solid india-rubber ring of the same shape . . .[11]

Furthermore, if two rings were linked, one through the other, that linkage would be indestructable, and they would travel as a linked pair to eternity. Could this be the explanation for the existence of the different sorts of atoms, the elements? The vortex ring would be the basic building block of matter, and by linking a few of them in different immutable ways, perhaps the different elements could be constructed.

Maxwell, was the scientific editor of the 9th edition of the *Encyclopaedia Britannica*, together with Thomas Huxley, and he wrote an article for the encyclopaedia on Atoms, in which he considered this theory. Its drawback is that it postulates an aether, and moreover one which to be a 'perfect' fluid cannot itself be atomic. But then the atoms produced are permanent and have structure. This last point was crucial: atoms were seen to emit spectral lines and so were assumed to have internal vibrations. If atoms were regarded as pure points they could not wobble. Theories allowing atoms to wobble usually imply they are made of something that can be pulled apart into smaller pieces. Vortex rings were unique in that they could vibrate but could *not* be destroyed. The final point in favour of the vortices was that there was a natural and necessary complication built into the system by the 'linkage' idea explaining why the atoms of different elements are different. 'The difficulties of this method are enormous, but the glory of surmounting them would be unique'.[12]

But Maxwell was quick to spot the possibility of taking a single vortex and knotting it together in different ways:

> Glenlair,
> Dalbeattie,
> Nov.13, 1867.
>
> Dear Tait,
>
> If you have any spare copies of your translation of Helmholtz on "Water Twists" I should be obliged if you could send me one.
>
> I set the Helmholtz dogma to the Senate House in '66, and got it very nearly done by some men, completely as to the calculation, nearly as to the interpretation.
>
> Thomson has set himself to spin the chains of destiny out of a fluid plenum as M. Scott set an eminent person to spin ropes from the sea sand, and I saw you had put your calculus in it too. May you both prosper and disentangle your formulae in proportion as you entangle your worbles. But I fear that the simplest indivisible whirl is either two embracing worbles or a worble embracing itself.
>
> For a simple worble may be easily split and the parts separated, but two embracing worbles preserve each other's solidarity thus

> though each may split into many, every one of the one set must embrace everyone of the other. So does a knotted one.

[13]

There is only one way of constructing a trefoil knot (or the 3rd order of knottiness), and only one way to make the 4th order knot that Maxwell draws, but after that the more twisted the vortex, the more ways there are of tying it. 47 ways for 9-fold knottiness, 123 for 10-fold. By 9-fold knottiness, Tait already had enough knots to explain all known elements, though they did not fall into any useful grouping. Although the vortex atom has now been discarded, this study of knottiness by Tait, inspired by Maxwell, is now highly regarded as one of the cornerstones of the modern mathematical theory of topology.

Largely through this work, they all developed a friendship and respect for 'the old Berlin Archimage', Helmholtz. Helmholtz was a brilliant mathematician, physicist and physiologist, a unique combination. They invited him to come to the B.A. meeting in Edinburgh in 1871 (as a good German, he first convinced them that merely because the French were losing the Franco-Prussian war, this did not mean that the French were in the right). Thomson made the first overtures, promising to take Helmholtz on his yacht, the *Lalla Rookh*. Then Tait added his blandishment:

> If Sir William be not ready, you might come with me to St. Andrew's, where my wife will be delighted to see you, and where you may learn (at its head-quarters) the mysteries of GOLF![14]

Tait was a golfing fanatic. His son Freddie was the British amateur champion and helped Tait with some research that certainly made him better known than any of his topology ever did. Tait showed that in driving a golf ball, backspin is inevitably given to the ball and that carries it farther than it would travel if unspun. There is a story, which is, unfortunately, untrue, that Tait had calculated the maximum possible distance a golf ball could be driven. Then up stepped Freddie, and knocked one with inadvertant backspin 250 yards down the fairway, and proved his father wrong.

Helmholtz duly went to St. Andrews and observed. Tait's biographer had the tact to leave Helmholtz' letter home to his wife in its original German in his work, but here in English is an unbiassed view of *das Golfspiel*:

> St. Andrews has a splendid bay, with fine sands which slope sharply up to the green links. The town itself is built on stony cliffs. There is a lively society of sea-side visitors, elegant ladies and children, and gentlemen in sporting costumes, who play golf. This is a kind of ball-game, which is played on the green sward with great vehemence by every male visitor, and by some of the ladies:—a sort of ball game in which the ball lies on the ground and is continuously struck by special clubs until it is driven, with the fewest possible blows, into a hole, marked by a flag, about an English mile distant. The entire round over which each party wanders amounts to about ten English miles. They drive the ball enormously far at each blow. Mr. Tait knows of nothing else here but golfing. I had to go out with him; my first strokes came off—after that I hit either the ground or the air. Tait is a peculiar sort of savage; lives here, as he says, only for his muscles, and it was not till to-day, Sunday, when he dared not play, and did not go to church either, that he could be brought to talk of rational matters.[15]

Helmholtz's own mathematical faculties obviously also enjoyed their holiday at St. Andrews.

If Tait managed eventually to convince Thomson of the respectability of atomic theory, one subject on which he failed miserably was that of Quaternions. These began in a growing need, stimulated largely by research into electricity and magnetism, to develop mathematical techniques suitable for handling quantities which had direction as well as magnitude, i.e. vectors.

Sir William Rowan Hamilton (the *other* Sir William Hamilton) was an Irish mathematician of genius. By the age of fourteen he was fluent in Latin, Greek, Hebrew, Italian, French, Arabic, Sanskrit and Persian. He was elected professor of astronomy in Dublin at the age of twenty-two while he was still an undergraduate. He was a compulsive poet and a friend of Wordsworth who gave him this sound advice:

> You send me showers of verses which I receive with much pleasure, as do we all: yet have we fears that this employment may seduce you from the path of science which you seem destined to tread with so much honor to yourself and profit to others. Again and again I must repeat that the composition of verse is infinitely more of an art than men are prepared to believe, and absolute success in it depends upon innumerable *minutiae* which it grieves me you should stoop to acquire a knowledge of . . . Again I do venture to submit to your consideration, whether the poetical parts of your nature would not find a field more favourable to their exercise in the regions of prose; not because those regions are humbler, but because they may be gracefully and profitably trod, with footsteps less careful and in measures less elaborate.[16]

He became a heavy drinker, then a total abstainer for two years, succumbed once more after being taunted for sticking to water at an astronomical party at Lord Rosse's (who was the possessor of the biggest telescope in the world, a forty-inch monster, housed in his castle in Ireland) and then, finally, Hamilton became a confirmed alcoholic. One Victorian biographer blamed this on 'the want of order that reigned in his home. He had no regular times for his meals but resorted to the sideboard when hunger compelled him. What more natural in such condition than that he should refresh himself with a quaff of that beverage for which Dublin is famous—porter labelled XXX?'

Hamilton also invented quaternions. He was a believer in moments of inspiration, as he later wrote to his son:

> On the 16th day of October, which happened to be a Monday, and Council day of the Royal Irish Academy, I was walking in to attend and preside, and your mother was walking with me along the Royal Canal, to which she had perhaps driven; and although she talked with me now and then, yet an undercurrent of thought was going on in my mind, which gave at last a result, whereof it is not too much to say that I felt at once the importance. An electric circuit seemed to close; and a spark flashed forth, the herald (as I foresaw immediately) of many long years to come of definitely directed thought and work, by myself if spared, and at all events on the part of others, if I should even be allowed to live long enough distinctly to communicate the discovery. Nor could I resist the impulse—unphilosophical as it may have been—to cut with a knife on a stone of Brougham Bridge, as we passed it, the fundamental formula with the symbols i, j, k; namely,
> $$i^2 = j^2 = k^2 = ijk = -1,$$
> which contains the solution of the problem.[17]

But even in his lifetime this momentous piece of graffiti had worn away.

Tait became interested in quaternions while at Cambridge, and when he went to Belfast he started to correspond with Hamilton. He became a convert, and a fervent one at that. The unbelievers claimed that quaternions were an empty language; pretty equations could be written in quaternion terminology, but only once the results were known; to derive those results, it was first necessary to revert to ordinary co-ordinate notation.

Thomson belonged to the ranks of the scoffers, which caused some degree of friction in the writing of T and T'. In 1901 Thomson wrote:

> We have had a thirty-eight year war over quaternions. He had been captured
> by the originality and extraordinary beauty of Hamilton's genius in this respect;
> and had accepted I believe definitely from Hamilton to take charge of
> quaternions after his death, which he has most loyally executed. Times without
> number I offered to let quaternions into Thomson and Tait if he could only
> show that in any case our work would be helped by their use. You will see that
> from beginning to end they were never introduced. [18]

Maxwell was a half convert:

> But try and do the 4ions. The unbelievers are rampant. They say "show me
> something done by 4nions which has not been done by old plans. At the best it
> must rank with abbreviated notations."
> You should reply to this, no doubt you will.
> But the virtue of the 4nions lies not so much as yet in solving hard questions,
> as in enabling us to see the meaning of the question and of its solution, instead of
> setting up the questions in x,y,z, sending it to the analytical engine, and when
> the solution is sent home translating it back from x y z so that it may appear as
> A, B, C to the vulgar. [19]

Maxwell with his pictorial vision was delighted with a shorthand technique
which enabled him to take in a problem at a glance. Even if quaternions were
merely an abbreviated notation, that was no reason to despise them. The
Mertonian schoolmen in fourteenth century Oxford (p.17) had run out of
steam largely because of poor notation. They discovered the mean speed
theorem for constant acceleration: $s = \frac{1}{2}(u+v)t$ but did not possess any algebra to
express it simply. Instead of the letter a for constant acceleration, they had to
write 'uniformly difform motion' and wield sentences rather than formulae.

Maxwell's concern for the language of science ran deeper than just notation.
To have clear and useful definitions was also a major help in research. In one
of his last articles, written for *The Electrician* in April 1879, he said:

> There is no more satisfactory evidence of the progress of a science than when
> its cultivators, having settled all their differences about the connexions of the
> phenomena proceed to reconstruct the definitions. Even in the most mature
> sciences, such as geometry and dynamics, the study of the definitions still leads
> original thinkers into new regions of investigation. [20]

Maxwell took the quaternion notation and tried to invest it with a
geometrical meaning, particularly the quaternion operator ∇ which is still used
in today's vector analysis.

Glenlair, Dalbeattie,
Nov.7, 1870

Dear Tait,

$$\nabla = i\frac{d}{dx} + j\frac{d}{dy} + k\frac{d}{dz}.$$

What do you call this? Atled?

I want to get a name or names for the result of it on scalar or vector functions
of the vector of a point.

Here are some rough hewn names. Will you like a good Divinity shape their
ends properly so as to make them stick?

(1) The result of ∇ applied to a scalar function might be called the slope of the function. Lamé would call it the differential parameter, but the thing itself is a vector, now slope is a vector word, whereas parameter has, to say the least, a scalar sound.

(2) If the original function is a vector then ∇ applied to it may give two parts. The scalar part I would call the Convergence of the vector function, and the vector part I would call the Twist of the vector function. Here the word twist has nothing to do with a screw or helix. If the word turn or version would do they would be better than twist, for twist suggests a screw. Twirl is free from the screw notion and is sufficiently racy. Perhaps it is too dynamical for pure mathematicians, so for Cayley's sake I might say Curl (after the fashion of Scroll). Hence the effect of ∇ on a scalar function is to give the slope of that scalar, and its effect on a vector function is to give the convergence and the twirl of that function. The result of ∇^2 applied to any function may be called the concentration of that function because it indicates the mode in which the value of the function at a point exceeds (in the Hamiltonian sense) the average value of the function in a little spherical surface drawn round it[21] ... What I want is to ascertain from you if there are any better names for these things, or if these names are inconsistent with anything in Quaternions, for I am unlearned in quaternion idioms and may make solecisms. I want phrases of this kind to make statements in electromagnetism and I do not wish to expose myself to the contempt of the initiated, or Quaternions to the scorn of the profane.[21]

In fact practically all these terms were adopted in principle, grad (short for gradient) was chosen instead of 'slope', though the idea is the same; 'convergence' has become divergence, through a mere reversal of sign; and for the vector part of ∇ applied to a vector the name 'curl' has stuck. The operation ∇^2 has stayed anonymous, unfortunately, since 'concentration' is a graphic and useful description of its effect.

The term 'atled' was not adopted for ∇ even if the symbol does look like an upside down delta. Instead they applied to W. Robertson Smith. He had begun his career as Tait's assistant at Edinburgh, while reading theology. There he had the unfortunate task of trying to persuade a particularly garrulous young student to concentrate on his physics—Robert Louis Stevenson. Robertson reported that he was lazy, idle and irritating (in Latin) and prevented his experimental partner, Marshall, from doing as well as he ought.[22] Stevenson dropped out and Marshall then went on to become Professor of Physics at Queen's University, Kingston, Ontario. But the experience may have dissuaded Robertson from physics, for after two years he left to become Professor of Hebrew in the Free Church College, Aberdeen. Suspected of being heterodox in his views on biblical history, he was later dismissed. Then he became the editor of the *Encyclopaedia Britannica*, which he ran most efficiently before moving back into academic life and becoming Professor of Arabic at Cambridge.

It patently required someone of Robertson Smith's erudition to do a Whewell for the Quaternionists, and he came up with the name Nabla—because ∇ is the shape of an eponymous Assyrian harp.

Tait as the inheritor of Hamilton's mantle was an ardent proselytizer for quaternions. Maxwell's approach was, as in everything, more whimsical:

Here is another question. May one plough with an ox and an ass together? The like of you may write everything and prove everything in pure 4nions, but in the transition period the bilingual method may help to introduce and explain the more perfect.[23]

He even poked fun at Tait's attitude in a referee's report for the Royal Society of Edinburgh:

> I beg leave to report that I consider the first two pages of Professor Tait's Paper on Orthogonal Isothermal Surfaces as deserving and requiring to be printed in the Transactions of the R.S.E. as a rare and valuable example of the manner of that Master in his Middle or Transition Period, previous to that remarkable condensation not to say coagulation of his style, which has rendered it impenetrable to all but the piercing intellect of the author in his best moments.[24]

Tait took that in good part, but he was annoyed when Maxwell attributed to Hamilton some work he had in fact done himself (it was concerned with another of Hamilton's great creations, the Hamilton characteristic function, the application of path integral methods to optical systems). Maxwell hastily had to apologise:

> O T' Total ignorance of H and imperfect remembrance of T' caused $\frac{dp}{dt}$ to suppose that H in his optical studies had made the statement in the form of a germ which T' hatched. I now perceive that T' sat on his own egg, but as his cackle about it was very subdued compared with some other incubators, I was not aware of its origin when I spoke to B.A.[25]

Normally, though, Maxwell's comments were accurate and incisive. When T and T' were preparing their book, they sent him the proofs, and received a particularly interesting comment about their definition of mass.

> 207. Matter is never perceived by the senses. According to Torricelli quoted by Berkeley 'Matter is nothing but an enchanted vase of Circe, fitted to receive Impulse and Energy, essences so subtle that nothing but the inmost nature of material bodies is able to contain them'. . .
>
> 208. Newton's statement is meant to distinguish matter from space or volume, not to explain either matter or density.
>
> "Def. The mass of a body is that factor by which we must multiply the velocity to get the momentum, and by which we must multiply the half square of the velocity to get its energy.
>
> You may place the two masses in a common balance (which proves their weights equal), you may then cause the whole machine to move up or down. If the arm of the balance moves parallel to itself the masses must also be equal.[26]

The first remark demonstrates Maxwell's philosophical carefulness, setting him off from T and T', and the second shows how clearly he distinguished between mass and weight. This led him on to some very interesting speculations about the theory of gravity, as he wrote to the astronomer William Huggins.

> . . . Any opinion as to the form in which the energy of gravitation exists in space is of great importance, and whoever can make his opinion probable will have made an enormous stride in physical speculation. The apparent universality of gravitation, and the equality of its effects on matter of all kinds are most remarkable facts, hitherto without exception; but they are purely experimental

facts, liable to be corrected by a single observed exception. We cannot conceive of matter with negative inertia or mass; but we see no way of accounting for the proportionality of gravitation to mass by any legitimate method of demonstration.[21]

Here is a clear realisation of the paradox whose reconsideration fifty years later by Einstein led to the General Theory of Relativity. Maxwell was interested at that time in the way the tails of comets streamed away from the Sun. Was it possible that comets were composed partly of gravitating matter and partly of levitating? Boiled off by the heat of the Sun, did their tails then stream away because they were repelled?

The overriding impression gained from Maxwell's correspondence, though, is of such overwhelming fertility of imagination that it continually overflowed into jokes, half-epic poems and quirkiness. The seriousness was always there, but coated in a layer of banter, as if Maxwell was always one step removed from the business at hand, lost in a reveries in some higher plane of existence.

The British Association met in Belfast in 1874. Maxwell attended, but Tait did not, and so Maxwell sent him a report of the activities. Maxwell was a little jaundiced in general with the B.A.

> Ye British Asses, who expect to hear
>   Ever some new thing,
> I've nothing new to tell, but what I fear,
>   May be a true thing. . . .[28]

In this particular year, Tyndall was in the presidential chair. Tyndall was an agnostic and evolutionist, and his presidential address inspired Maxwell for Tait's benefit:

> In the very beginnings of science, the parsons, who managed things then,
> Being handy with hammer and chisel, made gods in the likeness of men;
> Till Commerce arose, and at length some men of exceptional power
> Supplanted both demons and gods by the atoms, which last to this hour. . . .
>
> Then gathers the wave of emotion, then noble feelings arise,
> Till you all pass a resolution which takes every man by surprise.
> Thus the pure elementary atom, the unit of mass and of thought,
> By force of mere juxtaposition to life and sensation is brought;
> So, down through untold generations, transmission of structureless germs
> Enables our race to inherit the thoughts of beasts, fishes, and worms.
> We honour our fathers and mothers, grandfathers and grandmothers too;
> But how shall we honour the vista of ancestors now in our view?
> First, then, let us honour the atom, so lively, so wise, and so small;
> The atomists next let us praise, Epicurus, Lucretius, and all;
> Let us damn with faint praise Bishop Butler, in whom many atoms combined
> To form that remarkable structure, it pleased him to call—his mind.
> Last, praise we the noble body to which, for the time, we belong,
> Ere yet the swift whirl of the atoms has hurried us, ruthless, along,
> The British Association—like Leviathan worshipped by Hobbes,
> The incarnation of wisdom, built up of our witless nobs,
> Which will carry on endless discussions, when I, and probably you,
> Have melted in infinite azure—in English, till all is blue.[29]

Tait was delighted with the Ode, and after a round of golf, showed it to John Blackwood the publisher of Blackwood's Magazine. Blackwood made this cryptic note in his ledger:

> British Association was given to me in the club room at St. Andrews by Professor Tait, and I liked it so much that I said I would print it in the magazine if the author was so disposed. He was, but I know nothing further about him. . . .[30]

Maxwell was then the Cavendish Professor at Cambridge. Tait wrote to Maxwell for his permission, and received this reply:

> O T' not being professedly a Buffon the fugitive paroxysms of a Red Lion are not to be promiscuously stereotyped. The better ½ however desires me to make an exception in favour of Ebony [(Black Wood)]. . . .[31]

(The 'Red Lions' are a social club of the B.A.).

Having dealt with Tyndall, Maxwell continued in the same vein on Herbert Spencer, busy applying the laws of evolution to the development of civilisation:

> The ancients made enemies saved from the slaughter
> Into hewers of wood and drawers of water;
> We moderns, reversing arrangements so rude,
> Prefer ewers of water and drawers of wood.[32]

Finally he turned to his friend Professor Clifford, who read a paper 'On the General Equations of Chemical Decomposition':

> In Section B, Prof. W. K. Clifford read a paper on Chemical equations. The equation
>
> $$XX + LL = 2(XL)$$
>
> was the first selected. He observed that both the constituents of the left member were in the liquid state and that though the resultant might not be familiar to some members, he could warrant it 2XL. From an equation of similar form
>
> $$H_2 + Cl_2 = 2HCl$$
>
> he deduced by an easy transformation
>
> $$H^2 - 2HCl + Cl^2 = O,$$
>
> whence by extracting the square root
>
> $$H - Cl = O \text{ or } H = Cl,$$
>
> a result even more remarkable than that obtained by Sir B. C. Brodie.[33]

His private quirkiness did occasionally slip out into his review articles for *Nature*. In the review of the second edition of T and T', which Maxwell called the Archiepiscopal Treatise, because the Archbishops of York and Canterbury at that time were called Thomson and Tait (the Tait in question also being an old boy of Edinburgh Academy), Maxwell wrote this about the two authors' rigorous introduction of Lagrangian methods into the book.

> The two northern wizards were the first who, without compunction or dread, uttered in their mother tongue the true and proper names of those dynamical concepts which the magicians of old were wont to invoke only by the aid of muttered symbols and inarticulate equations. And now the feeblest among us can

repeat the words of power and take part in dynamical discussions which but a few years ago we should have left for our betters.[34]

That may have been slightly off-putting for T and T', but when Maxwell wanted to, his criticism could cut deep. He reviewed *Practical Physics, Molecular Physics and Sound* by Frederick Guthrie, Professor of Physics in the Normal School of Science, Kensington and gave it a thoroughly deserved trouncing (see p.49). Three weeks later, *Nature* carried a reply from Guthrie:

> Some well-meaning friend has composed and sent me a copy of the inclosed. There appear to be various opinions as to the authorship. It has even been suggested that Prof. Maxwell, with that sense of humour for which he is so esteemed, and with a pardonable love of mystification, is himself the author.
> February 24.   Fredk. Guthrie

Remonstrance to a Respected Daddie Anent His Loss of Temper Suggested by Prof. Clerk Maxwell's review of Guthrie's "Physics",

> WORRY, through duties Academic,
>   It might ha'e been
> That made ye write your last polemic
>   Sae unco keen:
>
> Or intellectual indigestion
>   O' mental meat,
> Striving in vain to solve some question
>   Fro' "Maxwell's Heat."
>
> Mayhap that mighty brain, in gliding
>   Fro' space tae space,
> Met wi' anither, an' collidin',
>   Not face tae face.
>
> But rather crookedly, in fallin'
>   Wi' gentle list,
> Gat what there is nae help fro' callin'
>   An ugly twist.
>
> If 'twas your "demon" led ye blindly,
>   Ye should na thank him,
> But gripe him by the lug and kindly
>   But soundly spank him.
>
> Sae, stern but patronising daddie!
>   Don't ta'e't amiss,
> If puir castigated laddie
>   Observes just this:—
>
> Ye've gat a braw new Lab'ratory
>   Wi' a' the gears,
> Fro' which, the warld is unco sorry,
>   'Maist naught appears.
>
> A weel-bred dog, yoursel' must feel,
>   Should seldom bark.
> Just put your fore paws tae the wheel,
>   An' do some Wark.

$$d\sqrt{\tfrac{m}{n}}.$$

The *nom de plume* here is taken from Maxwell's review of the book. It would be typical of Maxwell to apologise to Guthrie for the necessity of having been uncomplimentary by writing a self-depreciating poem.

Maxwell had felt strongly about Guthrie's book because it proclaimed itself as meant for schoolchildren. Maxwell believed that:

> In the popular treatise, whatever shreds of the science are allowed to appear, are exhibited in an exceedingly diffuse and attenuated form, apparently with the hope that the mental faculties of the reader, though they would reject any stronger food, may insensibly become saturated with scientific phraseology, provided it is diluted with a sufficient quantity of more familiar language. In this way, by simple reading the student may become possessed of the phrases of the science without having been put to the trouble of thinking a single thought about it. The loss implied in such an acquisition can be estimated only by those who have been compelled to unlearn a science that they might at length begin to learn it.
>
> The technical treatises do less harm, for no one ever reads them except under compulsion[36]

What was required was *good* scientific popularisation, to interest children in real science, and to give them a true feeling for the working of science without leaving them a mass of misconceptions it would take years to eradicate. Maxwell's own contributions to this were a couple of supposedly elementary books (*Heat*, and *Matter and Motion*) and his numerous articles for *Nature* and the *Encyclopaedia Britannica*.

One important piece of popularisation done during the Glenlair period was Maxwell's work on the Cambridge Tripos. He was a moderator or examiner in 1866, 1867, and 1870, and took the opportunity to revolutionise the syllabus, feeling that beforehand the Wranglers had 'spent their time upon mathematical trifles and problems, so-called, barren alike of practical results and scientific interest'[37]. He introduced questions on electricity and magnetism and recent research (see p.79, 83) and tried, as he had done at King's, to make the problems relate to the real world.

In this academic venture he was completely successful, but he was not when he applied to become Principal of St. Andrews in 1868 on James Forbes' retirement. Lewis Campbell was the Professor Greek there at the time, and states that Maxwell decided not to apply for the job, but Professor Domb has recently shown that Campbell was being somewhat discreet in his account.

Maxwell was a popular choice amongst the faculty of St. Andrews, but he initially felt that 'my proper line is in working not in governing, still less in reigning and letting others govern'. Also the severe East Wind in those exposed coastal areas was a definite drawback. On 3 November 1868 he wrote to Campbell saying he had decided against applying, but Professor Domb has found a letter in the Royal Institution Archives, from 7 November, saying he had now decided to stand. Probably he had visited St. Andrews on 5 November and been persuaded. On 9 November he wrote to Thomson asking for a testimonial:

> When I last wrote I had not been at St. Andrews. I went last week, and have gone in for the Principalship. If you can certify my having been industrious etc.

since 1856, or if you can tell me what scientific men are conservative or still better if you can use any influence yourself in my favour pray do so.[38]

His asking Thomson to name *conservative* scientists is highly significant. The appointment was in the hands of the Home Secretary, and Disraeli's Conservatives were in power. On 11 November, however, the date of Forbes' official retirement, the Government fell. Gladstone and the Liberal party won the following election, and the new Home Secretary promptly appointed the Professor of Latin, Shairp, to the job. Perhaps Maxwell had identified himself too strongly with the Conservatives.

Having received this snub, Maxwell retreated back into his Glenlair shell, and it was only with some difficulty that he was prised out again by Cambridge University to become the first Cavendish Professor of Experimental Physics in 1871.

CHAPTER 10

Campbell and Garnett trace the beginnings of Maxwell's work on gas theory back to the discovery of the planet Neptune, one of the epic tales of scientific detection.

The planets of the solar system are all much less massive than the Sun, and so to a first approximation, their orbits are wholly determined by the Sun's gravitational field. To a second approximation the gravitational forces they exert on each other cause those orbits to wobble. The observations of the wobble of the then outer-most known planet, Uranus (discovered by William Herschel in 1781) failed to match these calculations. Some thought this showed the inverse square law to be wrong, others that there was some resisting medium in outer space affecting Uranus' movement. But in 1841 it occurred to John Couch Adams, then an undergraduate at St John's College, Cambridge, that these strange wobbles might be explained by the existence of another, hitherto undiscovered planet lying further out than Uranus. This inspiration was only the start of his work; in those days before the computer, it took him four years of longhand mathematics to calculate that everything fitted: there *was* a new planet, and he was able to work out where it would be seen. George Biddel Airy, the Astronomer Royal, completely ignored the prediction.

Meanwhile, the same hypothesis had independently occurred to the young French scientist Urbain Leverrier, and in 1846 he wrote to the Observatory in Berlin. On the same day that they received his letter they trained their telescopes to the spot indicated and so it was German astronomers who gained the glory of discovering Neptune on 23 September 1846. The new addition to the solar family was headline news, naturally enough, and one of the scientific lectures that Maxwell was taken to hear as a schoolboy was given by Mr. Nichol, Professor Astronomy at Glasgow, on the discovery of Neptune.[1]

In academic circles there was a certain amount of chauvinist bickering about the discovery before the credit was allowed to devolve equally on both Adams and Leverrier. Promptly, some of Adams' admirers from his Cambridge college collected money together to found a prestigious biennial mathematical prize in his honour. The subject set for the 1855 prize was 'The Motions of Saturn's Rings'—a tough problem on the stability of gravitational orbits squarely in the Adams tradition. Why do Saturn's rings stay up, spinning round the planet's equator like great hula-hoops rather than come spattering down on the planet's surface?

The scale and the magnitude of the problem gripped Maxwell and in an opening paragraph to his prize-winning essay, after he too had taken four years to do the necessary calculations in between his duties at Aberdeen, he wrote:

> When we contemplate the Rings from a purely scientific point of view, they become the most remarkable bodies in the heavens, except, perhaps, those still less *useful* bodies—the spiral nebulae. When we have actually seen that great arch swung over the equator of the planet without any visible connexion, we cannot bring our minds to rest. We cannot simply admit that such is the case, and describe it as one of the observed facts in nature, not admitting or requiring explantion. We must either explain its motion on the principles of mechanics, or admit that, in the Saturnian realms, there can be motion regulated by laws which we are unable to explain.[2]

The classic work on orbital stability is *La Méchanique Céleste* by the French mathematician Pierre Simon Laplace. It was there for the first time that the problem of the mutual interactions of the planets was solved, thus forming the basis of Adams' work. There also Laplace had a look at Saturn's rings, and showed that if the rings were rigid and uniform (which they seemed to be to astronomers on Earth) then they could not possibly be stable: they would have to crash sooner or later. With the intuition of mathematical genius, Laplace suggested (but could not *prove*) that if the rings were *non*-uniform that is their mass was loaded differently at different points around the circumference, then they could be stable.

Maxwell first addressed himself to this problem, and found that Laplace had indeed been correct—but only just. Saturn's rings, if rigid, would be stable only if between 81.6% and 82.8% of their mass was concentrated at a single arbitrary point on the circumference with the rest evenly distributed. It seemed most unlikely that Saturn's rings could have acquired this extraordinarily delicate dynamical wheel balancing by 'accident'. In any case, though his mathematical intuition was superb, Laplace's physical intuition did not perhaps quite match, the stresses involved in holding a rigid ring together against the tremendous centrifugal shear of Saturn's gravity, would smear out any known material like warm butter.

Mathematically, the rigid ring was the simplest case because it held together as a unit. Now Maxwell had to look at non-rigid rings, and take into account that they could vibrate and wobble internally. He had to discover whether these internal wave motions could be stable. One possibility was a fluid ring rotating like a wheel (i.e. with all its mass having the same angular velocity) but with internal flexibility. Making use of the analytical technique of Fourier analysis (an abstract mathematical technique, far from the geometical methods 'normally' associated with Maxwell), he was able to show that the internal wave motion must inevitably break up such a fluid wheel into a chain of drops.

The only possibility left to Maxwell was that Saturn's rings were made up of swarms of small particles, moving in independent orbits round the planet, with different speeds at different radii. They look solid to us on Earth because of their great distance. A complete analysis of the motion of such a swarm is impossible, but Maxwell broke the problem down into a simpler form, which he could tackle by extending the Fourier analysis he had used earlier. He considered a ring of equally spaced, equally massive particles. The differential equation for their motion is of the fourth order, which means that for a vibration of any given wavelength, there will be four separate frequencies of vibration. Every one of those different waves for every possible wavelength must be individually stable if the ring as a whole is to be stable. In fact this is not as tough a criterion as it may sound; if Saturn's mass, relative to the mass of the ring-particles is sufficiently great, then the ring will be stable—and Saturn is a very massive planet.

To illustrate some of the sorts of wave motion that could exist in the ring, Maxwell had a model built by a local craftsman, Ramage, 'for the edification of sensible image worshippers',[3] a sly dig at his friend Thomson who was also one of the examiners for the prize. Thomson loved to invent mechanical models (only in his mind) of his theories:

> It seems to me that the test of 'Do we or do we not understand a particular subject in physics?' is 'Can we make a mechanical model of it?'[4*]

Maxwell's miniature hand-cranked orrery, of brass, wood and ivory is still kept in the Cavendish Laboratory in Cambridge.

All through the work, in fact, Maxwell had kept in touch with Thomson.

14 Nov 1857

> Dear Thomson
>
> I suppose, after a busy vacation, you are very busy in the beginning of the session. Nevertheless you must allow me to report progress on Saturn's rings, so that if you have any remarks to make on my results, I may have the benefit of them. I have already reported on the rigid rings. I got your investigation of it, but I have been entirely occupied with the non-solid ring since so that I have only *read* it, not *worked* it yet.[5]

That was slightly unethical perhaps, since Thomson was one of the prize examiners. The note by Thomson referred to in the letter was published at the end of Maxwell's final paper.

Of course, a regular, bead necklace of satelites looks nothing like Saturn's rings, but it was a start. The real rings could be regarded as being built up from lots of Maxwell's rings each with a slightly different radius. But an extra complication is how these separate rings of satellites would interact with each other. Maxwell considered the mutual interaction of two rings, and found that for certain ratios of their radii, and certain orders of the wave, the wave induced in the one ring by the other would resonate, its amplitude should grow slowly but uncontrollably and eventually disrupt the ring.

This was as far as Maxwell's analysis could go, quantitatively, the *real* problem, how to attack the mutual interaction of a broad ring of independent particles moving in individual orbits, was still open, as he wrote to Thomson:

> The general case of a fortuitous concourse of atoms each having its own orbit and eccentricity is a subject above my powers at present, but if you can give me any hint as to the point of attack I will go at it.[6]

* Duhem makes great fun of Thomson's view, although in essence it is very close to Einstein's use of *gedanken* experiments.

But Thomson could not help, and Maxwell could make only qualitative progress with the problem. Collisions between the particles of different rings would result in an effective frictional force between adjacent rings. Add this to the possible resonance between rings of quite different radii, and the result should be a general smearing out; the outside rings should move even farther away, and the inner rings inwards. As Maxwell put it, more graphically, in another letter to Thomson:

> What shall we say to a great stratum of rubbish jostling and jumbling round Saturn without hope of rest or agreement in itself, till it falls piecemeal and grinds a fiery ring round Saturn's equator, leaving a wide tract of lava with dust and blocks on each side and the western side of every hill buttered with hot rocks . . . As for the men of Saturn I should recommend them to go by tunnel when they cross the "line".[7]

Sadly, Maxwell refrained from including that description in his final paper.

Maxwell's essay won the prize. His general conclusions that the rings of Saturn are clouds of independent particles was later shown to be correct by observation. It is delightful to note, though, that the detailed understanding of the rings has not advanced all that much since Maxwell's time. The reason why the rings have not all 'buttered' Saturn's equator is still not understood, and the flypast of Saturn by Voyager 1 in 1980, almost a 150th birthday present to Maxwell, showed that one of the rings had 'spokes' and another was braided—a phenomenon till then thought to be impossible.

Maxwell's letters to Thomson have an extra significance. 'What will be the effect of collisions?' he asked. Since treating each particle individually is obviously impossible, how do you deal with a 'mass of rubbish jostling and jumbling round'? The ground was prepared for his next great piece of work, for his first paper on the theory of gases, where he turned his attention from the vastness of the Universe to the smallness of the atom, and considered what was happening inside a gas if you assumed it was made of millions of atoms jostling and jumbling against each other.

Campbell and Garnett suggest the connection was direct, but Maxwell himself acknowledged another, equally important stimulus, in the papers of the German physicist Rudolf Clausius. The basic notions of the kinetic theory of gases, went back a long way. The Greeks had the idea of atoms, of course, and the atomic model underpins Newton's mechanical Universe, but the first serious attempt to make use of the *motion* of atoms came from one LeSage. If you imagine the whole of space to be filled with rapidly moving invisible atoms of gravity, then a body sitting in the middle will have these atoms bumping into it from all directions. The net force would therefore be zero. If another body is put nearby, however, it will cast a 'shadow' around it and shield the first body from collisions in its direction. And vice versa. The forces on the two bodies are now unbalanced: there are more collisions on their far sides than on the sides which face, and they shield each other. It is easy to show that this force should naively go as the inverse square of the distance, but it is equally easy to kick great holes in the theory, as Maxwell did in an article on 'Atoms' he wrote for the *Encyclopaedia Britannica* (the shadowing effect should exactly

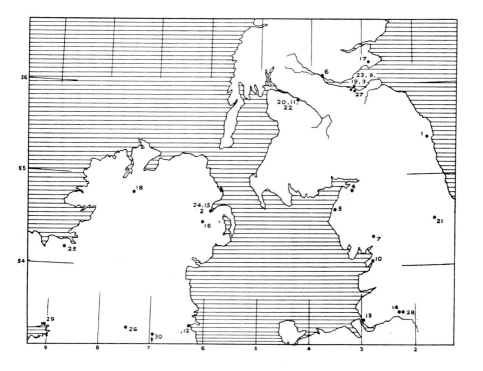

1. *Map of the birthplaces of 19th century scientists (not exhaustive)*

1 Airy	16 Larmor
2 Andrews	17 Leslie
3 Balfour Stewart	18 MacCullagh
4 W H Bragg	19 Clerk Maxwell
5 Dalton	20 Melvil
6 Dewar	21 Peacock
7 Eddington	22 Ramsay
8 Fitzgerald	23 Rankine
9 Forbes	24 Osborne Reynolds
10 Frankland	25 Stokes
11 Graham	26 Stoney
12 Hamilton	27 Tait
13 Jevons	28 J J Thomson
14 Joule	29 Townsend
15 William Thomson	30 Tyndall

2. *Maxwell as a boy from one of his cousin Jemima's watercolours. (Cavendish Laboratory, University of Cambridge).*

3. Maxwell: the young scientist at Cambridge. He is holding the "improved" colour top. (Cambridge University Library).

4. Professor Maxwell. (Peterhouse College, Cambridge).

5. *Professor Maxwell from a drawing made when he was the Cavendish Professor. The strain of establishing the laboratory (amongst other things) seems to have turned his hair grey. (Cavendish Laboratory)*

6. *Professor and Mrs Maxwell plus a shaggy dog, posed in front of a painted backdrop to simulate a suitably romantic Highland atmosphere. (Cavendish Laboratory).*

7. *Maxwell's dynamical top, with 9 adjusting screws and a colour disc on top. (Cavendish Laboratory).*

8. The model made by Ramage in Aberdeen to demonstrate Saturn's rings. A handle is turned at the rear and the little ivory balls at the front can wiggle in two of the different wave-modes of a single ring. (Cavendish Laboratory).

9. Apparatus to measure the viscosity of gases. Its heat jacket has been lost, but at least that allows one to see the fixed and rotating vanes inside. (Cavendish Laboratory).

10. The balance arm wired up to compare electromagnetic against electrostatic units of charge, which ratio, at Maxwell's hands, gave the speed of light. (Cavendish Laboratory).

11. Apparatus with which Maxwell and MacAlister repeated and improved Cavendish's experiment to test the validity of the inverse square law of electrostatic repulsion. They found it true to one part in 21,600. (Cavendish Laboratory).

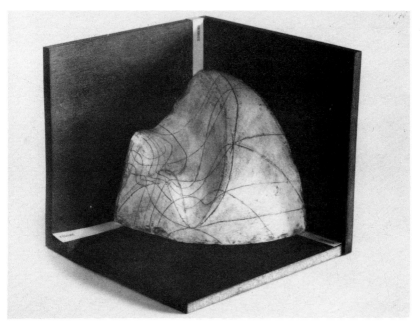

12. *Maxwell's model of Gibbs' thermodynamic surface for water. (Cavendish Laboratory).*

13. *The zoetrope. Maxwell's peep-show apparatus. The cartoon strip on the inside is his work, and shows tumbling acrobats. (Cavendish Laboratory).*

14. *The Cavendish Laboratory a short while after Maxwell's death, with students doing a variety of demonstration experiments. There are two women present! (Cavendish Laboratory).*

15. *Glenlair. The building that was Maxwell's country seat: less grandiose than his father had planned.*

balance the ricochet effect, particles bouncing off one particle and onto the other).

The first serious attempt to apply the idea of atomic impacts to material bodies came from Daniel Bernoulli, who stated that the pressure exerted by a gas could be interpreted as the gas being made up of molecules, colliding with and ricocheting from the walls of their container. In the middle of the nineteenth century, a number of physicists, Herapath, Joule, Krönig and Clausius took Bernoulli's intuition and started making it into mathematical physics.

There is one name missing from that list, that of John James Waterston. Waterston was brought up, like Faraday, in the strict tenets of the Sandemanian Church, but unlike Faraday, he lapsed. Watserston was a Scot and went to the University of Edinburgh; to earn a living he trained as a surveyor, before going to teach ballistics to the East India Company's naval cadets in Bombay.

As a nineteen year old student, he had produced his own explanation of gravity, like Le Sage filling space with particles, though his were long cylinders whirling around like drum-majorettes' batons. Then in December 1845 he submitted from India a paper to the Royal Society, where he derived for the first time many of the basic results of the atomic theory of gases. The paper is a curious one, a lot of the ideas are good, but the mathematics is crazy. Sadly, the Royal Society referees did not draw even this distinction:

> the paper is nothing but nonsense, unfit even for reading before the Society.[8]

Unfortunately by then it already had been read, and by the rules of the Society, a read paper that was not to be published was not returned to the author (so he could try again elsewhere) but sat mouldering in the Society's archives.

Waterston made a few attempts to let the world know of the existence of the paper, but these were unsuccessful, and it was not till 1892 that Lord Rayleigh discovered it and had it belatedly published. Among its original results were the equipartition theorem and the ratio of the specific heats of gases at constant pressure and at constant volume, two topics which Maxwell was later to explore. Waterston's intuition in both was correct, though he made glaring mathematical errors. Waterston had the misfortune to be both ahead of his time and unknown, as Rayleigh noted.

> The omission to publish . . . at the time was a misfortune, which probably retarded the development of the subject by ten or fifteen years . . . At any time since 1860 reference would have been made to Maxwell, and it cannot be doubted that he would have at once recommended that everything possible should be done to atone for the original failure of appreciation.[9]

Rayleigh is also reported to have said that if the scientists who developed kinetic theory had been billiards players, the subject would also have developed quicker, since billiards is a good mechanical, pictorial analogy for the way molecules collide in gas theory. Now for a young student in Cambridge, billiards was regarded as the ultimate in depravity; once a man was hooked on billiards he was thought to be good for little else. Even rowing was regarded with less academic disfavour, so there is a degree of irony in Rayleigh's remark.

Waterston, the lapsed Sandemanian out in India, had plenty of time on his hands and did become highly proficient at billiards—one wonders how much of his paper owes its genesis to the sinful game. Of course, billiards is played with only a few balls, and, at the hands of an expert, bears little resemblance to a random gas. Snooker or pool played by a beginner is a much better analogy.

Maxwell became interested in these ideas after reading a translation of some of Clausius' work, published in the *Philosophical Magazine* in 1859. The major advance made there was an explanation of the diffusion of gases. Theory showed that the pressure in a gas was proportional to the mean square of the speed of the molecules (actually $p = \frac{1}{3} \rho \bar{v}^2$ where ρ is the density) but to make theory and observation agree, the velocity of the molecules had to be extremely high, of the order of several hundreds of meters a second.

Why, then, if a bottle of pungent, volatile chemical was opened in one corner of a room, did its molecules not rush out almost immediately to assault the nostrils of an observer in the other corner of the room? Molecules, in fact, take quite a long time to diffuse across a room where there are no draughts. Clausius explained that by molecular collisions: the molecules do not fly straight across the room, but cannon into other molecules in the way. They do move fast, but are constantly jostled into new paths by such collisions. To get from bottle to nose, a molecule has to travel enormously further than just the straight-line path, and takes a long time to do this. Maxwell himself expressed the paradox thus:

> If you go 17 miles per minute and take a totally new course 1,700,000,000 times in a second where will you be in an hour!?[10]

In his work, Clausius had made the assumption that at any one temperature, all the molecules had the same speed. Maxwell immediately realised that this could not be so. Even if all the molecules started off thus, collisions would inevitably knock some molecules off faster and some slower. It was the same problem as the spreading, colliding rings of Saturn, but this time Maxwell found the answer.

First, he had to overcome a philosophical block. Atoms, the Scottish philosophers thought, might be useful hypothetical objects to build up a picture of phenomena, but since they could not be detected, one had to be very careful about saying that they actually existed; where did atoms end and phrenology, phlogiston and the aether begin. One of them (Brougham) once made this memorable attack on idle speculation:

> A mere theory . . . is the unmanly and unfruitful pleasure of a boyish and purient imagination, or the gratification of a corrupted and depraved appetite. . . .[11]

Waterston, brought up on the same philosophical doubts as Maxwell, had cautiously hedged his bets.

> Whether gases do consist of such minute elastic projectiles or not, it seems worth while to inquire into the physical attributes of media so constituted, and to see

what analogy they bear to the elegant and symmetrical laws of aeriform bodies.[12]

Maxwell felt these philosophical doubts equally strongly: it might be interesting to show that well-established laws could be derived quite simply from the general features of a mathematical model, but he was proposing to explain how the individual molecular velocities in the gas were distributed. Disbelief in the existence of molecules can be suspended only so far. At the end of May, 1859, Maxwell wrote to Stokes:

> I do not know how far such speculations may be found to agree with facts, even if they do not it is well to know that Clausius' (or rather Herapath's) theory is wrong, and at any rate as I found myself able and willing to deduce the laws of motion of systems of particles acting on each other only by impact, I have done so as an exercise in mechanics. Now do you think there is any so complete a refutation of this theory of gases as would make it absurd to investigate it further so as to found arguments upon measurements of strictly "molecular" quantities before we know whether there be any molecules?[13]

In fact Maxwell had already found the perfect way out of his dilemma. Until then, the molecular theory had merely expressed old laws in a new language, but Maxwell had found a new and totally unexpected prediction. It was testable and therefore exposed molecular theory to Hamilton's criterion of falsifiability. If the prediction survived experiment, then molecular theory would at the very least, have to be taken seriously.

His letter to Stokes immediately continued with this prediction, that the viscosity of a gas is independent of the pressure. When a body moves through a gas it is, on average, hit harder by the molecules on its front than on its rear, and so it loses forward momentum. This frictional effect of the gas is its viscosity. It would be natural to think that the denser is the gas, the treaclier it would be, the greater would be this effect. But while the rate of gas-body collisions increases with the density of the gas, so too does the rate of gas-gas collisions, and so the rate at which atoms escape from the surface of the body, leaking away its momentum is also decreased. The increase in the number of gas-body impacts is exactly balanced by the increase in intermolecular collisions, and the net drag remains constant.

This is a splendid prediction: it is surprising, relatively straightforward to verify, but yet insensitive to the fine details of molecular theory (Maxwell asked Stokes whether, by chance, he just happened to have already done the experiments).

Encouraged by this prediction, Maxwell gave a lecture on the whole of his work on gas theory at the B.A. in Aberdeen in September 1859, publishing the results in the *Philosophical Magazine* the year after.[14] Important though his viscosity formula was, the really vital result contained in the paper was the prediction that the speeds of the molecules should *not* all be constant as Clausius had suggested, but they should, instead, be spread out in a particular way, known ever since as the Maxwell distribution. Equally, the argument Maxwell used has baffled physicists ever since.

He said, let the distribution of velocities in the X direction be described by a function f, so that the number of molecules with velocity x is f(x). Now this distribution, according to Maxwell, cannot affect the Y distribution of velocities which must take the same form, so one gets a factor f(y) and another f(z). But the X, Y, and Z axes are arbitrarily chosen, and so the number of molecules with velocity components x, y, and z can depend only on the total velocity $v = \sqrt{x^2 + y^2 + z^2}$. So $f(x) \; f(y) \; f(z) = \phi(\sqrt{x^2 + y^2 + z^2})$. The solution to this equation is, $f(x) = C \exp(-\alpha x^2)$. And this, with a suitable normalising factor, gives the Maxwell distribution.

It is worth considering this argument. There is an assumption that the distribution in mutually perpendicular directions must be independent. Why? It is at least worth considering the possibility, say, that if the velocity y is very large, then there might be a diminished possibility that x and z should simultaneously be large also.

Even more striking, however, is the manner of the formula's derivation. It is supposed to represent the distribution of velocities to which a gas settles down, no matter how it starts off, once the molecules are given the chance to buffet and jostle each other. Yet the formula makes no mention of the collisions of particles at all. The functional relation is conjured out of thin air. Maxwell's argument seems to have no physics in it at all. In fact Maxwell's argument is formally almost identical to one John Herschel produced in an article for the *Edinburgh Review* of 1850, to derive the error law in statistics: the way chance scatters results round a mean value.

Professor Everitt has shown that the wording of a letter written probably in June 1850 to Campbell (containing the portentous phrase 'the true Logic for this world is the Calculus of Probabilities') indicates that Maxwell had read Herschel's review. Nine years later, faced with the problem of the distribution law, Maxwell dredged up Herschel's argument from the recesses of his memory. He wrote to Stokes:

> Of course my particles have not all the same velocity, but the velocities are distributed according to the same formula as the errors are distributed in the theory of least squares. [15]

It never will be known which came first in Maxwell's mind, the argument or the result, but one strongly suspects the latter: the formula has a nice logical shape, it relates to a well-known result, and he adapted the derivation of that result for the purpose. Here was the answer to the problem he had faced with Saturn's rings; faced with an enormous number of independent particles, the only thing to do is to ignore most of their individual properties, and instead treat them statistically, by which method some of their gross characteristics can be calculated. Maxwell's distribution law can never say what is the velocity of any individual particle, just that some values are more likely than others. In Maxwell's unpublished papers there is, in fact, an unsuccessful attempt to apply the formula directly to Saturn's rings.

Maxwell's *Phil. Mag.* paper comes in three parts and contains a multitude of results. Using his diffusion law he derived formulae for the so-called transport phenomena: diffusion, considering the way a molecule jostles about carrying its 'smell' with it, viscosity, the way it carries momentum (mentioned above) and finally thermal conductivity, the way a molecule carries kinetic energy, identified with heat, with it. The three are all closely related.

Maxwell applied his formula to various test cases, or at least attempted to. He calculated that the thermal conductivity of air is 10,000,000 times less than that of copper. Clausius pointed out that Maxwell had forgotten to convert from kilogrammes to pounds and from hours to seconds. The actual factor is 7,000, so when Maxwell later wrote that numerical factors are a nuisance (p.80) he spoke from bitter experience. Another mistake which Clausius pointed out was that Maxwell had used his distribution law derived for a constant temperature to calculate thermal conductivities where there was necessarily a temperature gradient across the gas. Clausius tried to correct this, but he insisted on sticking with his model of uniform velocities.

Of a quite different order were Maxwell's mistakes while attempting to derive the equipartition theorem (unaware that Waterston had already proposed this), which states that in a gas energy is divided equally between all possible modes of motion: there is on average as much X-kinetic energy as Y, and the rotational energy round any axis should also have the same value as the kinetic energy. First Maxwell attempted to show that when molecules of diferent masses were mixed, they would, after some collisions end up with equal average kinetic energies: their mean square velocities should be inversely proportional to their masses.

However, he considered only molecules actually travelling with exactly the respective mean square velocities of the two species, which was just the hypthesis for which Maxwell had criticised Clausius, and furthermore he allowed only perpendicular collisions. Within these highly restrictive conditions, he showed that there was a tendency for the energies of molecules of different masses to equalise after impact, but the falseness of his premise is shown by the fact that if the molecules had the same mass (i.e. mixing two portions of the same gas each at a different temperature), then by his calculations the temperatures should equalise instantaneously.

In the third part of the paper is an even more bizarre piece of work attempting to prove another aspect of the theorem, the equality of rotational and kinetic energies. The method was once again the Clausius trick of replacing all velocities by their mean value, which Maxwell further compounded by getting his statistical techniques all wrong.

Not surprisingly, Maxwell was able to prove the result that he wanted in both cases. The interesting thing is that this result is perfectly correct, Maxwell intuitively knew the answer, and trying to work backwards from that and forwards from the Clausius hypothesis, hoped the middle would take care of itself. This method of guessing a symmetrical result (the distribution formula, the equipartition theorem) and then setting about establishing a basis for believing it was a long way from the logical, inductive approach to science advocated by Bacon and the Scottish philosophers.

Maxwell was so convinced of this equipartition theorem that he immediately used it to check the ratio of the specific heat of a gas at constant pressure to its value at constant volume, a ratio known in the trade as γ. If one ignores the rotational energy of a gas, then in heating it through 1° Centigrade at constant volume, the energy goes straight into increasing the average kinetic energy of each molecule, by a constant amount along each axis, defined as k/2. Thus the specific heat of a molecule at constant volume, the amount of energy needed to heat a single molecule through one degree Centigrade, is 3k/2. If one heats the same gas at constant pressure, the same amount of heat goes into molecular motion, but the gas also expands. If one imagines the gas constrained inside a piston, then as it expands, extra energy is used up in pushing back the piston (i.e. the atmosphere).

The German physicist Mayer had shown that the work done on the piston per degree centigrade is k, so adding these contributions together, the specific heat per molecule at constant pressure is 3k/2 + k = 5k/2, and the ratio of the two values is (5k/2)/(3k/2) or 5/3. Experiments had been done to measure this ratio, and though fairly crude, they showed quite convincingly that the value for most gases was not 1.666 but nearer 1.4 (Waterston had also been the first to try the calculation, but had, typically, made a mistake, obtaining the value 4/3 rather than 5/3 he should have got, and so he was quite happy with his result).

If one assumed that an atom was merely a point, a centre of a field of force and no more, then it was reasonable to assume that it had no internal substance itself to rotate. But Maxwell was concerned with molecules, not atoms. A molecule was envisaged as the simplest conglomerate of point atoms that could exist freely; it might consist of many atoms locked by their mutual interactions into a rigid structure. So even if the atoms were point-like the molecules were not, they could bump into each other obliquely and so rotate.

Maxwell now applied his analysis to γ. At constant volume there should be 3k/2 for kinetic energy, and, by equipartition, 3k/2 for rotational energy. At constant pressure, 3k/2 + 3k/2 + k. The ratio should now be (3k/2 + 3k/2 + k)/ (3k/2 + 3k/2) = 4/3. This is the same value as Waterston had been delighted with, but unfortunately in the intervening years experimental accuracy had progressed considerably and the value Maxwell had to fit was 1.408 not 1.333.

He was naturally very upset at this; the value seemed to be a robust prediction of the equipartition theorem, which in itself was a straightforward deduction from the basic principles of the molecular theory of gases. Maxwell was so confident of its truth that he said:

> This result of the dynamical theory, being at variance with experiment, overturns the whole hypothesis, however satisfactory the other results may be.[16]

Before abandoning the theory, however, he decided to test his other robust prediction, that viscosity be independent of pressure. The experiment took a long time, he moved to London, had to settle in there, had to prepare his courses and he had his B.A. Committee work. He then ran into some design difficulties and his first attempt at the experiment (1863) had to be abandoned because the apparatus was not airtight—the great bugbear of low-pressure work. "It is the case of 'the little rift within the luting' which bothers us all in vacuum experiments"[17], he confided to Tait.

It was not till 1865 that he was happy with his design (see Fig X), though even then there were still occasional mishaps, but ever the good scientist, Maxwell took advantage of one of these to study the fracture patterns generated in ½-inch glass plate.

> I made an erroneous estimate by rule of thumb as to the strength of a glass plate ½-inch thick, in consequence of which when exposed to a pressure of ¾ atmosphere, it succumbed with a stunning implosion and sent me a month back.[18]

The apparatus, now kept in the Cavendish Laboratory, is of heroic design. Glass discs on a rotating axis are interleaved with fixed glass discs: the friction of the gas between the fixed and moving discs gives a measure of the viscosity. The moving discs are suspended on a steel wire with a magnet attached to the bottom. An external electromagnet can be used to twist the axis and then release it. The discs swing slowly round, as the steel wire unwinds, and the viscosity is derived from the way this torsional rotation is damped out. The motion is measured by a tiny mirror attached to the wire, which gives a light but very sensitive detector of small twists of the wire. The pressure inside the apparatus is controlled by a pump.

While he was at it, Maxwell also decided to test the dependence of viscosity on temperature, and this involved placing a jacket around the vessel into which steam or cold water could be passed. Campbell's account of Mrs. Maxwell's heroic Turkish bath efforts in the cause of science has already been dismissed (p 73). One charming note is added in Maxwell's paper that to insulate the tin steam jacket round the apparatus, he wrapped it up well in blankets and put a feather cushion on the top.

He completed his experiments in London by the end of the 1865 academic year, and took the raw data with him to Scotland for the recess. Unfortunately he forgot to take any log tables and so had the tedious job of reducing the data by hand, but the results fully confirmed his prediction: for a large range of pressures, the viscosity was indeed constant.

The way the viscosity depended on temperature, however, came as a considerable surprise, as he wrote to the Glasgow-educated physicist Thomas Graham, who following in Newton's footsteps was also the Master of the Mint!

> The friction is the same for all densities, but increases with the temperature, apparently in the same proportion as the air expands. I expected it would be as the square root of the absolute temperature but I think I am wrong. These results agree with yours.[19]

The billiard ball model gave the clear prediction that as well as being independent of pressure, the viscosity should also increase as the square root of the absolute temperature, but Maxwell found that the ratio of the viscosity at the highest temperature he could attain (185°F) to the lowest (51°F— none of the experiments using ice are recorded) gave a value 1.2624. The ratio of the temperatures on the Absolute scale is 1.2605. The closeness of the two numbers led him to believe that viscosity is directly proportional to absolute temperature, a result which, when he had recovered from his initial surprise, he found fitted in very well with his intuitive personal philosophy of science.

The simplicity of the other known laws relating to gases warrants us in concluding that the viscosity is really proportional to the temperature, measured from the absolute zero of the air-thermometer.[20]

This is a fascinating insight into his mind: simplicity was for him a significant physical criterion; he expected God to create the world in a way that he could understand as rational and symmetric. A rational God would obviously provide linear laws in Nature; later in Cambridge he tested and confirmed the linearity of Ohm's law to 1 part in 10^{12}. Here he already knew that Boyle's law, Charles' law and all the other gas laws were linear relations, so he felt it just that the viscosity law should be so also. In fact, he was wrong, and later admitted in a further paper that more refined experiments had shown that the viscosity law is not linear at all.

Maxwell's apparatus was later taken to Cambridge and his experiment repeated. The values he had quoted for viscosities were then found to be rather large, which was tracked down to Maxwell calculating his results for a different (and wrong) value for the radius of his rotating discs to the one he used in the description of his apparatus. It is as well that Thomson did not know this when he referred the paper for the Royal Society. He was so impressed that he said such experiments were of National Importance, that the Royal Society should fund a Laboratory where they could be carried out (the genesis of the National Physical Laboratory?) and that he, or Maxwell should run it.[21]

The different status of Maxwell's two predictions about viscosity is very important. He had found a good simple physical explanation why viscosity should be independent of pressure, which should be valid for any molecular model of a gas. It was vital that experiment should confirm this result, and it did. The prediction of its dependence on the square root of the temperature was quite different; there was no general reason why this should be so, it was a specific feature of the particular model Maxwell had used—the billiard ball model. When experiment showed the result to be wrong, it merely indicated that the billiard ball model was inadequate, and another approach was in order.

Maxwell now took the apparently surprising step of looking at different forms of intermolecular force laws. In his work on electromagnetism (next chapter) one of Maxwell's guiding lights was a dislike for action-at-a-distance, the way gravity and electricity seemed to spring invisibly across a vacuum and act on distant bodies with an inverse square law, with the mechanism of the force remaining entirely mysterious. Maxwell usually liked to embody his mathematics in some physical picture of the force, to find a physical mechanism or medium by which the force could be transmitted—as Le Sage and Waterston had attempted in their models of gravity.

His billiard ball model would appear to be a perfect mechanical model by Maxwell's criteria, with nothing but direct contact interactions between the molecules, no action-at-a-distance here. But the arguments against direct contact forces were as strong as those against action at a distance. If two *rigid* particles collide, they must change their directions of motion instantaneously, which implies an infinite force acting for an infinitesimal time, which is an absurdity. Therefore particles cannot be rigid, they must have structure and be made of smaller components, but, by the same argument, these too could not be smaller rigid spheres themselves. To stop the infinite regression, Roger Boscovich, a Serbian Jesuit, proposed that the ultimate particles of Nature had no size; they were truly point particles, which had mass, and a field of force around them. Contact forces were discarded and action at a distance brought back and as far as he (and the Scottish philosophers) was concerned, the absurdity of infinite forces had been exchanged for the unpleasantness of action at a distance, albeit distances of 10^{-8}cm, the size of an atom.

It was to such force laws that Maxwell turned in his great paper of 1866 to explain the effect of temperature on viscosity[22]. Billiard ball collisions were out, and Maxwell had to try something new. What he said, in effect, was 'let us assume that molecules push on each other with a force that goes as r^{-n} and see what happens'. In his 1860 paper he had explicitly recognised that the billiard ball model is equivalent to a very strong short-range repulsion (i.e. making n very large in the force law). This was now exploded so other values of n were to be tried.

In a head-on collision between two billiard ball molecules both rebound equally; the same thing happens for any law of repulsion, although now they will rebound without actually touching. For glancing incidence, however, the effects are no longer the same for different values of n. Billiard balls just pass each other by, with no interaction, but repulsive molecules still affect each other. In general, the amount of their mutual deflection depends on their relative velocities, the faster they are moving, the less time there is for the forces to act, and so the smaller the deflection. On the other hand, the faster they move, the greater the number of collisions they suffer per second. In general then, the effect on the force is compounded with that of the velocity distribution in an extremely complex and intractable way.

For one very special force law, however, the mathematics simplifies enormously. When n = 5, an inverse fifth power repulsion, the increase in the collision rate with velocity *exactly* balances the decrease in deflection per collision, and the relative velocity of the molecules drops out from the equations entirely, and it is now possible to solve them. Maxwell did this, calculated the viscosity—and found it depended linearly on the temperature. He looked no further.

Maxwell would probably have regarded this as another manifestation of simplicity and God's largesse to physicists. In fact he had made a mistake. The fortuitous cancellation of the velocity is true only for two-molecule collisions. The simultaneous collision of three or more molecules must also be taken into account, and these many-body collisions disrupt the simplicity of Maxwell's result for the viscosity, but for the relatively low pressures in which Maxwell was interested, these higher order terms make only a small contribution.

Maxwell used his new model to recalculate the transport phenomena equations of his earlier paper, and to solve the equations in a number of idealised cases. One point he rather glossed over is the ratio of the specific heats. If the billiard ball model of 1.333 'overturned the theory', then this model, for 'point' molecules has to give 1.666 which is even worse—as Thomson who referred the paper again strongly emphasised and Maxwell himself admitted[23]. By avoiding physical contact between his molecules, there was no way that they could interchange kinetic and rotational energy, and so the rotational component of the energy has to be 'frozen' out of γ. From the start, therefore, Maxwell knew that this model had to be incomplete.

He also had another look at the velocity distribution formula admitting that the assumption of the mutual independence of velocities in the three perpendicular directions 'may appear precarious'. This time he did consider actual collisions, but the derivation seems scarcely less magical. He assumes there is some velocity distribution $f(v)$. Then the number of collisions between molecules travelling with speeds v_1 and v_2 must depend on $f(v_1)f(v_2)$. Let their speeds after collision be v_1' and v_2'. Now consider a collision between 2 molecules whose *initial* speeds are v_1' and v_2', this must go as $f(v_1')f(v_2')$.

At equilibrium the number going from $v_1 v_2$ to $v_1' v_2'$ must exactly balance those going in the reverse direction, and with the constants of proportionality cancelled $f(v_1)f(v_2) = f(v_1')f(v_2')$. Now if $v_1 v_2$ scatter to $v_1' v_2'$, there is an extra constraint on the process, the conservation of kinetic energy or, $\frac{1}{2}mv_1^2 + \frac{1}{2}mv_2^2 = \frac{1}{2}m(v_1')^2 + \frac{1}{2}m(v_2')^2$. Together, these two equations once more give $f = N \exp(-\alpha v^2)$, but it is unlikely that anyone dubious about the first derivation would be converted by this one.

The status and significance of his distribution law may not even have been clear to Maxwell himself at the time. One of the problems he considered in the first draft of the paper was the dependence of temperature on height in a vertical column of gas. As he frankly admitted in his final version of the paper, his first attempt at a solution to this problem went wrong because of a fairly straightforward mathematical error, and led to a solution where the temperature dropped off so rapidly with height that the column became unstable: perpetual currents of air would result. Maxwell wrote a private letter to Thomson about this saying he was worried. Thomson in his referee's report for the Royal Society recalled that he too had been worried—as well he might, since the gas column appeared to break both the 1st and 2nd laws of thermodynamics which Thomson had done so much to establish.

Maxwell found a mistake in his calculation and produced an improved solution where the temperature now increased with height. This was at least mechanically stable. Further reflection, however, showed that this too broke the 2nd law, and further checking showed that this time he had made a simple arithmetical error. The final solution was that the temperature is independent of the gravity and the 2nd law stayed intact.

A number of rather subtle points emerged from this little academic comedy. Firstly Maxwell submitted, yet Thomson did not immediately and categorically veto, a paper defying the 2nd law of thermodynamics. No paper would be so leniently treated today. On the other hand, when the paper was finally corrected, compliance with the 2nd law was seen to depend rather directly on the specific mathematical properties of Maxwell's distribution formula. Clausius' distribution law, for example, would inevitably defy the 2nd law.

The full implication of the connection became clear only with the invention of the Maxwell demon, a delightful beast who has tormented theoretical physicists ever since.

He first appeared in a letter to Tait, on 11 December 1867:

> Now conceive a finite being who knows the paths and velocities of all the molecules by simple inspection but who can do no work except to open and close a hole in the diaphragm by means of a slide without mass . . .[24]

Maxwell proceeded theoretically to set up a container with a diaphragm in the middle separating a hot gas on one side from a cold gas on the other. But in the cold gas, by Maxwell's formula there should still be *some* fast molecules, and in the hot gas, *some* slow ones. Maxwell's finite being, or demon as he became universally known, works his slide so that the fast molecules from the cold side pass to the hot side, and slow molecules from the hot side pass the other way:

> . . . that is the hot system has got hotter and the cold colder and yet no work has been done, only the intelligence of a very observant and neat fingered being has been employed.

Thus, Maxwell continues, if we were sufficiently nimble fingered we could break the 2nd law 'Only we can't, not being clever enough'.*

This notion intrigued Tait who promptly referred it on to Thomson for a higher opinion. Both T and T′ were convinced. The demon has led a checkered existence since then, and has in large part led to the current development of information theory. The generally held opinion now is that the demon cannot work; to see the fast and slow molecules arriving he would have to receive light from them, and, by the precepts of quantum mechanics, bouncing a photon of light off a molecule alters its velocity, so the demon could no longer tell if it was fast or slow.

But what the demon did show was that the 2nd law is actually a statistical law.

> The second law of thermodynamics has the same degree of truth as the statement that if you throw a tumblerful of water into the sea, you cannot get the same tumblerful of water out again.[26]

The demon could *not* work for Clausius' molecules, but the range and uncertainty of velocities in Maxwell's distribution gave him room for manoeuvre. But the statistical method of dealing with molecules is not merely a statement of our necessary inability to deal simultaneously with billions and billions of molecules, it is actually a necessity. The gravitational column and thus the 2nd law *depend* on the Maxwell distribution. The distribution is

---

* The nearest thing to a demon is the remarkable trickshot at pool played by Buster Keaton in the film *Sherlock Junior:* He hits the cue ball into a seemingly random pack, pocketing them all but one. It was not faked, but it did take Keaton whose grasp of mechanics must have rivalled Newton's a whole fifteen minutes to prearrange the balls on their proper spots so they would all travel precisely to their pockets.[25]

statistical, then so is the 2nd law, and with it the evolution of the Universe itself, since it is the 2nd law which is time's arrow, giving direction to the flow of time.

Gone was the clockwork certainty of the Newtonian Universe, and replacing it a world where the pulse of the clock behind the works itself became a gamble. The philosophical and religious implications did not escape Maxwell. In a clockwork universe there was no room for free will, everything was preordained. In the new universe there was more freedom. The discovery of the law of conservation of energy had knocked out the idea of the direct action of the will on the world: the muscles must act in accord with the principles of physics not the soul. Maxwell's answer was to say the soul could act like the driver of a steam engine, by operating a lever or a switch he could divert the power of the shuddering juggernaut in any direction he desired. In physical terms, the complexities involved in the interactions of large numbers of objects result in unstable situations, where the system is so precariously balanced (like a boulder on top of a hill whose support frost has eroded away) that the least nudge could push it in any desired direction. If the initial conditions were known with absolute precision such situations could not arise, but invariably there is some statistical uncertainty about the initial conditions and so these instabilities do occur. So when later on Clausius and other German physicists tried to derive the 2nd law as a purely dynamical phenomenon from Hamiltonian mechanics, paying no regard to its statistical character, Maxwell was very properly scornful.

It was, however, another German physicist, Ludwig Boltzmann, who became Maxwell's (often critical) apostle. He was converted by Maxwell's first paper and spent the next thirty years developing and refining Maxwell's ideas, and establishing statistical mechanics (as the subject developed) on a firm footing. One of his achievements was to include the external forces directly in the derivation of the distribution law. The answer to the column of gas in a gravitational field then comes out trivially. In 1873, in a spate of correspondence in *Nature*, Guthrie assailed Maxwell on this very point. Maxwell knew that Boltzmann's answer was better than his own, but even he had difficulty with Boltzmann's turgid style, as he wrote to Tait:

> By the study of Boltzmann I have become unable to understand him. He could not understand me on account of my shortness and his length was and is an equal stumbling block to me. Hence I am very much inclined to join the glorious company of supplanters and to put the whole business in about 6 lines.[27]

Maxwell rederived Boltzmann's result in a single page of *Nature*.

At the back of his mind however, remained the problem of γ:

> Still one phenomenon goes against that theory—the relation between specific heat at constant pressure and at constant volume, which is in air = 1.408 while it ought to be 1.333.[28]

Boltzmann found what at first sight seemed a neat solution to the problem. He assumed that 'air' molecules are made of just two point atoms locked into a rigid dumbell shape by the interatomic forces. If looked at from the side, such a molecule would have physical size, but if looked at end on, it would not: it was not unreasonable that it might not be possible to set it rotating about its own axis, though it would be able to rotate about the two perpendicular ones. From both the specific heat at constant pressure and at constant volume, one partition of the energy is removed, and the ratio is thereby reduced from 8:6 to 7:5, which at 1.400 is not a million miles from 1.408.

This seemed an impressive tribute to Maxwell's theory, but he himself was not seduced by it, for spectroscopists led by Stokes had discovered that there are discrete spectral lines in the light emitted by hot substances: that molecules when heated gave off light at precise and characteristic wavelengths (as well as a general non-specific glow: the difference between a sodium street lamp and an electric bulb). The light seen in the laboratory had exactly the same wavelength as light from those same molecules existing inside stars, which indicated to Maxwell that all molecules were mass-produced to the same specifications, by God. Even more exciting was the realisation that these spectral lines must be telling us vital new information about the internal structure of molecules. The lines were inherent to molecules, they could depend only on the molecules themselves, and so had to specify something, as yet unfathomed, about the way the parts of the molecules vibrated against each other. Each molecule had to have its own inner clockwork. As Maxwell wrote, in the first flush of this discovery:

> An intelligent student armed with the calculus and the spectroscope can hardly fail to discover some important facts about the internal constitution of a molecule.[29]

In this context, however, the lines were a decided nuisance. They implied some sort of internal vibration in the molecules: if they were agitated by heat, some internal clock pendulum was set swinging, ticking away with a precise period and radiating light at a precise frequency. But the equipartition theorem said that energy should be fed equally into *all* possible modes of oscillation and movement of the molecule. This internal vibration must also claim its equal share of the energy and must inevitably bring the value of γ back down, to lower than the unacceptable 1.333.

Boltzmann tried to wriggle out of the conundrum via the aether, by calling into play some 'mutual action between the molecules and the aetherial medium which surrounds them'. Maxwell wryly noted, in a lecture to the Chemical Society, 'I am afraid, however, that if we call in the help of this medium, we shall only increase the calculated specific heat, which is already too great'.[30] Maxwell's conclusion was that, 'something essential to the complete statement of the physical theory of molecular encounters must have hitherto escaped us'.[31]

That something was quantum theory, which showed some fifty years later just how at low temperatures, the different channels for the disposal of energy were 'frozen out'. At high temperatures the molecules do vibrate. Lower this temperature and effectively the molecule completely stops vibrating, and this degree of freedom for energy disposal is frozen out of the equipartition; since no energy goes into or out of the vibrational mode when the temperature changes, vibrations can now play no part in the molecule's specific heat. At lower temperatures still, the rotation of a diatomic molecule about its axis is also effectively frozen out, so Boltzmann was also correct.

It is worth noting, however, how cautious Maxwell was: he was not lulled into rushing into acceptance of his own ideas—even when in Boltzmann's hands they seemed to give agreement with experiment.

If this exemplifies Maxwell's philosophic detachment and caution, another passage from the same 1876 Chemical Society lecture demonstrates equally well his other gift of inspired intuition, the passage where he produced the Maxwell construction in thermodynamics.

It was well established experimentally that real gases do not obey the ideal gas law PV = RT exactly. For high pressures and/or low temperatures they deviate. Eventually they liquify, and the deviation becomes extreme. The Dutch physicist, Johannes van der Waals had tried to explain such deviations in terms of intermolecular forces and the finite size of molecules. For Maxwell this was a mixture of the two analogies he had used in his papers on gas theory, van der Walls wanted billiard ball molecules with force. In Maxwell's terms, van der Waals wanted a long range attraction superimposed on a short range repulsion (equivalent to the 'size' of the molecule). But Maxwell could deal with such a short-range force in his own terms, with the other forces, not as an *ad hoc* molecular size. Maxwell's review of van der Waals' paper for *Nature* carried his own calculations which gives a rather different result from that van der Waals' had obtained. Nevertheless, Maxwell said,

> . . . his attack on this difficult question is so able and so brave, that it cannot fail to give a notable impulse to molecular science. It has certainly directed the attention of more than one inquirer to the study of the Low-Dutch language in which it is written.[32]

This must be an oblique reference to himself since Campbell confides, the only language he had any difficulty in mastering was Dutch'.[33] Maxwell set van der Waals' formula as a problem in the Tripos within a couple of years of its publication.

Van der Waals' formula works very well phenomenologically until a gas starts to liquify. Since it is a simple continuous formula it cannot hope to deal with the onset of liquefaction, when inside the container there are simultaneously two phases present (both gas and liquid), and at that point if the pressure is increased only slightly all the gas converts to liquid, that is, there is an enormous change in the volume. The van der Waals' equation cannot hope to tackle this discontinuity.

Maxwell proposed a method of dealing with the problem. He started with the van der Waals' curve for a gas at some specific temperature which has a maximum and a minimum.

As you climb up from low pressure (a) you reach somewhere the liquefaction point (A). Then instead of the gas continuing to follow the curve, it liquefies at that pressure, its volume changes dramatically, and it breaks away from the curve sharply to move across horizontally to pick up the continuous curve on the far side (C), where it now is entirely liquid.

Maxwell now applied the 2nd law of thermodynamics, and in the best traditions of the subject, he considered how a heat engine would work going from A to C straight across and then back along the helter skelter of the curve. If the shaded areas were not equal, he said, the 2nd law would be broken. This is a pretty argument, displaying all Maxwell's feel for neatness and symmetry. Unfortunately it is also wrong, the helter skelter path involves unstable states to which classical thermodynamical arguments do not apply. It was nearly a century later before Maxwell's result was shown to be correct, by a valid argument.

It was Boltzmann again who inspired Maxwell to write one of the two great papers on gas theory he produced in the last year of his life. It was, basically, yet another derivation of the velocity distribution, but this time with a broader scope, to deal with energy rather than velocity distribution and it was couched in a very general, powerfully mathematical Hamiltonian formalism (this time his proof seems to be valid!). The power of the method is shown by the fact that the equipartition theorem just drops out as a by-product of the proof. Maxwell also carefully pointed out that collisions between the molecules have also disappeared (again) from the proof. Maxwell had the confidence to insist (correctly), that the distribution theorem should not just be applicable to gases, but that it was a general law of nature.

The opening paragraph of the paper is characteristic of Maxwell.

> Dr. Ludwig Boltzmann, in his "Studien über das Gleichgewicht der lebendigen Kraft zwischen bewegten materiellen Punkten" (*Sitzb. d. k. Akad. Wien*, Bd LVIII., 8 Oct, 1868), has devoted his third section to the general solution of the problem of the equilibrium of kinetic energy among a finite number of material points. His method of treatment is ingenious, and, as far as I can see, satisfactory, but I think that a problem of such primary importance in molecular science ought to be scrutinized and examined on every side, so that as many persons as possible may be enabled to follow the demonstration, and to know on what assumptions it rests. This is more especially necessary when the assumptions relate to the degree of irregularity to be expected in the motion of a system whose motion is not completely known.[34]

Not many would have been able to follow Maxwell's mathematics, but this willingness to entertain a multitude of approaches was typical. In a presidential address to the B.A. he once said (hearking back precisely to Common Sense views (p.44)):

> There are men who, when any relation or law, however complex, is put before them in a symbolical form, can grasp its full meaning as a relation among abstract quantities. Such men sometimes treat with indifference the further statement that quantities actually exist in nature which fulfil this relation. The mental image of the concrete reality seems rather to disturb than to assist their contemplations.

There are others who feel more enjoyment in following geometrical forms, which they draw on paper, or build up in the empty space before them.

Others, again, are not content unless they can project their whole physical energies into the scene which they conjure up. They learn at what a rate the planets rush through space, and they experience a delightful feeling of exhilaration. They calculate the forces with which the heavenly bodies pull at one another, and they feel their own muscles straining with the effort.

To such men momentum, energy, mass are not abstract expressions of the results of scientific inquiry. They are words of power, which stir their souls like the memories of childhood.

For the sake of persons of these different types, scientific truth should be presented in different forms, and should be regarded as equally scientific, whether it appears in the robust form and the vivid colouring of a physical illustration, or in the tenuity and paleness of a symbolical expression.[35]

Maxwell ended this 1879 paper by applying the theoretical principles he had developed to the case of a rotating tube of gas. He showed that there would be density gradients in the tube, which would be different for the different species of molecule in the gas. Here was developed the basic idea of the centrifuge, which is a vitally important device today in the separation and purification of chemicals.

Maxwell's other 1879 paper, was inspired by the radiometer. This is the 'perpetual motion' device seen in the windows of curio-shops, made of an evacuated glass bulb containing a paddle with its blades silvered on one side and darkened on the other. Expose it to light and it rotates. When this device was invented in the 1870s by William Crookes, it raised enormous interest and excitement, it seemed to reveal a new force of nature, for no-one could understand initially why it should go round. Maxwell attended a Royal Society soirée, where Crookes demonstrated the machine, and reported to Tait that he found it much more interesting than the then current vogue for poltergeists.

I saw Crookes experiments at the R.S.Soi . . . Ree. They whip spirits all to pieces. A candle at 3 inches acts on a pith dish as promptly as a magnet on a compass needle. No time for air currents and the force is far greater than the weight of all the air left in the vessel. Attraction by a bit of ice very lively. All this at the best obtainable vacuum. At 12mm pressure no effects at 760 mm pressure reverse effects but not nearly so prompt or even so strong as in a vacuum.[36]

Maxwell was the obvious person to referee Crookes' paper, and enthusiastically endorsed publication,

[it is] of such transcending scientific value that the record of the steps by which the discoverer was led to it becomes worthy of a permanent place in the history of science.[37]

In his work on electromagnetism, Maxwell had shown that light should exert a pressure, but this pressure was far too low to explain Crookes' result; but Maxwell was quite happy to see his theory overthrown.

In my treatise on Electricity, Arts 792 and 793, I have pointed out a probable repulsive action of radiation. But the effects observed by Mr. Crookes seem to indicate forces of much larger value.[37]

Crookes produced a stream of careful experiments and hundreds of other scientists joined in. It was not long before Tait and Dewar in Edinburgh and Schuster then in Manchester, performed convincing experiments to show that the effect actually depended on the small amount of gas left behind in the radiometer bulb, not on any new force of nature.

Maxwell, as the conscientious referee of Crookes' papers, overcame his initial enthusiasm to report wisely on the next paper that its importance would depend on the eventual explanation of the radiometer's action. By the time Crookes wrote his final radiometer paper (1878) Maxwell's report read:

> The labour, skill and ingenuity displayed in these experiments are marvellous and many of the individual results are valuable but I think the improvement of the methods of making and preserving vaccum is likely to be still more valuable to science than any of the actual experiments.[38]

And this is true, chasing the red herring of the radiometer force had led physics in general and Crookes in particular to improve vacuum technology, with dramatic scientific (and economic) consequences for the world. Hermann Bondi attributed the bulk of the advances made in the twentieth century to the vaccum pump.

Nonetheless, popular interest in the radiometer was so great that it attracted Queen Victoria's attention, and Maxwell was deemed a suitable person to demonstrated the toy to Her Majesty. He wrote to his uncle, Robert Cay, to describe the occasion but failed to say whether or not she was amused.

> . . . I was sent for to London, to be ready to explain to the Queen why Otto von Guericke devoted himself to the discovery of nothing, and to show her two hemispheres in which he kept it, and the pictures of the 16 horses who could not separate the hemispheres, and how after 200 years W. Crookes has come much nearer to nothing and has sealed it up in a glass globe for public inspection. Her Majesty however let us off very easily and did not make much ado about nothing, as she had much heavy work cut out for her all the rest of the day . . .[39]

Although the effect was now clearly attributed to the interaction of gas molecules with light and the radiometer vane, there was no consensus as to what this interaction was, and this led Maxwell to become interested in boundary effects in gas theory, what happens at gas-solid interfaces.

His paper was presented to the Royal Society in April 1878, his conclusion being that for a static situation, the force would depend on the second spatial derivative of the temperature $\partial^2\theta/\partial x^2$. Unfortunately a straightforward mathematical consequence of this fact is that an equilibrium temperature distribution is soon reached, where there is *no* resultant force. So how does the radiometer work? Maxwell said the mechanism had to be *slipping*, a phenomenon discovered by Helmholtz. When a temperature gradient exists along a solid surface immersed in a liquid or a gas, the fluid will tend to slide along that surface and exert a force on it. But Maxwell left his original paper with this as a bare qualitative statement: it was the only possible explanation he could think of, but he attempted no calculation of the effect.

It was almost inevitable that Stokes should now ask Thomson to referee the paper for the Royal Society, and as usual Thomson's comments were highly

illuminating, leading to a major theoretical advance. Thomson was enthusi-
astic about the paper, but was unhappy about Maxwell's rather bald
qualitative conclusion:

> §11, 12, 13 are of exceedingly great importance and interest. It is probable that
> Prof. Stokes may have known the principle before but infinitely improbable that
> any one else did. It is desirable that some slight additional information should be
> given to help the reader.[40]

Thomson dropped the pretence of anonymity and the next time he met
Maxwell suggested a method of developing the idea of slipping, by considering
an idealised surface consisting of gills, or sinusoidal or pyramidical hillocks.
These were well-behaved mathematically and it might be possible to work out
from the way individual molecules collided with them, how the gas as a whole
might slip.

Maxwell took up the challenge in an Appendix added in May 1879. His idea
was that there were two extremes of behaviour which are both readily
calculable. One is that molecules striking a surface are reflected precisely the
way light is reflected from a mirror. The other possibility is that they are
absorbed, they stick to the surface and are later evaporated off (in Thomson's
model, this corresponds to a surface with rows of deep veins on it, in which the
molecules get trapped and jostle about for a long time before they can escape
again). Any real surface must lie somewhere between these two extremes, and
so he parametrised a surface by f, the fraction of the incoming molecules to be
evaporated, and therefore $1-f$ reflected. He could now quantitatively calculate
the slippage in terms of f, and the important subject of gas-surface interactions
was born.

There is a multiple irony about this. First, Maxwell was trying to investigate
the radiometer, but as Thomson pointed out in his referee's report, he seemed
unaware of the work of Tait and Dewar, which had already showed that the
radiometer worked at very low pressures, below the limits where Maxwell's gas
theory approximations would break down. Maxwell's calculations apply to a
higher pressure regime than is found inside the bulb of a radiometer.

Secondly he was led to consider slippage because he thought that at
equilibrium, the ordinary steady-state pressures would equalise and give no
resultant force on the apparatus. The calculation is, admittedly, a very tricky
one, since the temperature gradients affect the molecular velocity distributions
which are used to calculate the temperature gradients. Perhaps it is not
altogether surprising therefore that Maxwell made a mistake here. In their
standard text on the theory of non-uniform gases, Chapman and Cowling
derived the correct mathematical expression for the effect, which includes an
extra term that Maxwell had overlooked. The result is that slipping is *not*
necessary to drive the radiometer even in the relatively high-pressure regime
Maxwell was considering; even without air currents and slippage, the
radiometer would go.[41] Nonetheless, Maxwell's was an original and pioneering
piece of physics.

Between Maxwell's first presentation of his ideas to the Royal Society in
April 1878 and the Appendix in May 1879, a rival had appeared on the scene.

Osborne Reynolds submitted a paper to the R.S.,which, among other things, included an explanation of the slippage effect. Maxwell and Thomson were asked to referee it.

This put Maxwell in an awkward position; here was a paper on a topic on which he was himself working. Displaying his usual interest in the pyschology of discovery, Maxwell felt that the experimental part of the paper should be published as it stood.

> I do *not* think it would be desirable to have a new edition of the experimental part, revised by the light of his maturer knowledge, because I think that such a faithful record of the process of sound scientific thought is more valuable than a textbook, however well digested.[42]

But in the theoretical section of the paper he noted some serious errors, and also passed on Thomson's helpful suggestion on how to treat surfaces mathematically. Nonetheless he thought Reynolds' conclusions were basically correct, and the paper should be published 'after the author has had the opportunity to make certain improvements in it'.

Reynolds had made claims in the paper that he alone had *proved* that molecules had a finite size. Maxwell took up this point, and showed with his usual caution, that because a scale-length could be associated with a gas, this did not necessarily mean that that length was 'the size' of a molecule. The other referee, Thomson, was far more vehement in his criticism. He quoted chapter and verse in his own work and that of others to show calculations of the size of atoms. But he found the implied sneer to Maxwell in the paper even more distasteful.

> I do think that the frame of mind of a writer who writes on the dimensional properties of gases without quoting either Clausius or Maxwell or Loschschmidt or Stoney is not suitable for the preparation of a paper for the Royal Society . . .[43]

Reynolds' only reference to Maxwell was to say he had got the whole thing wrong; now his reply was conciliatory.

> I have avoided further reference to Maxwell's paper because I did not feel it incumbent on me to point out what appears to me to be a weak point in Maxwell's reasoning. Considering what an authority Maxwell is you will understand my being very unwilling to believe that I had found an oversight in his work, and how after trying all ways . . . I could make nothing else of it I would rather mistrust myself and avoid the subject.[44]

In turn, Reynolds became very annoyed when Maxwell's final paper was published. Maxwell had been in a difficult position. He *had* had privileged access to Reynolds' paper, which did overlap with his own. He decided to go ahead and publish the Appendix, paying tribute to the 'referee' who had suggested he pursue the subject, and paying even fuller tribute to Osborne Reynolds' 'admirable work':

> This phenomenon [(thermal slipping)] was discovered entirely by him. He was the first to point out that a phenomenon of this kind was a necessary consequence of the Kinetic Theory of Gases.[45]

He continued that he had reconsidered the subject only after reading Reynolds' work and that his own method 'is, in some respects, better than that adopted by Professor Reynolds'. He had bent over backwards to give credit to Reynolds: he had become interested in slippage through the work of the Germans, and through Thomson's suggestion, even before he received Reynolds' paper. Nonetheless, it was undeniable that because the referees' reports (mostly Thomson's in fact) had delayed the publication of Reynolds' work, Maxwell had brought out an overlapping paper ahead of him. Reynolds was very angry and wrote to Stokes to complain, particularly that Maxwell had cast aspersions on his as yet unpublished paper. Stokes' typed reply was frigid:

> I may as well tell you in the first instance that Professor Maxwell is lying in a dying state. I don't feel certain that Dr. Paget has given up the last ray of hope, but I should think most probably he has. I imagine it is only a question of days or possibly weeks, but hardly that. . . . I must frankly say that there appears to me to be a sort of air of irritation in what you have written which I do not think the very courteous way in which Maxwell has spoken of your paper in his own justifies.[46]

That Maxwell was able to produce this final paper while gnawed by the incessant pain of cancer of the abdomen is perhaps not the least of his achievements.

Maxwell's work on gas theory reveals more about his intuitive style of research than does any other part of this oeuvre. 'He is a genius', said his great German contemporary, Kirchhoff, 'but one has to check his calculations before one can accept them'[47]. His ideas were always splendid, but his derivations and calculations not always so. Modesty forbade Maxwell to pass comments on his own especial talents, indeed he once wrote to Tait about the Maxwell construction on the van der Waals' curve, 'That I did not do it [(before)] shows my invincible stupidity', but he was able wholeheartedly to recognise his own carelessness:

> Clifton is quite right about Gordon having used a wrong formula. . . . I cannot be sure how far I am responsible for Gordon's formula. I am quite capable of writing a fancy formula and not finding it out till I come to work it.[48]

But this fades into insignificance in comparison with his remarkable achievements. Here is a heartfelt (though slightly inaccurate) tribute from Boltzmann, Maxwell's rival in so much of this work.

> A mathematician will recognise Cauchy, Gauss, Jacobi, Helmholtz, after reading a few pages, just as musicians recognise, from the first few bars, Mozart, Beethoven or Schubert. Perfect elegance of expression belongs to the French, though it is occasionally combined with some weakness in the construction of the conclusions; the greatest dramatic vigour to the English, and above all to Maxwell. Who does not know his dynamical theory of gases? At first the Variations of the Velocities are developed majestically, then from one side enter the Equations of State; from the other the Equations of Motion in a Central Field; ever higher sweeps the chaos of Formulae; suddenly are heard the four words: 'put n = 5.' The evil spirit (the relative velocity of two molecules) vanishes and the dominating figure in the bass is suddenly silent; that which had seemed insuperable being overcome as by a magic stroke.[49]

Maxwell's magic was his intuition: he had the confidence of genius in believing in the simple neat formulae his intuition suggested. But because he was working backwards from those results so much of the time, it is not altogether surprising that his contemporaries were mystified.

CHAPTER 11

If there is one discovery with which Maxwell's name will be indelibly associated, it is the formulation of the set of equations for the electromagnetic field universally known as Maxwell's equations. If his work on the theory of gases shows him at his most inspirational, then his work on electromagnetism reveals most clearly the philosophical side of his genius.

In his first paper, at Cambridge, he constructed a hydrodynamical model of the aether, which embodied the known laws of magnetism and the theoretical insights of Michael Faraday in 'pictorial' form. Faraday's experimental work (and that of others) had shown, however, that electricity and magnetism interconnected very strongly: moving electricity produced a magnetic field and moving magnetism produced electric fields. This first model therefore had to be incomplete. Maxwell threw away this model, and constructed an entirely new model of the aether, much more sophisticated, with gears and ball bearings sprouting out all over space. Not only did it embody *all* the known experimental laws of the subject within the Faraday framework, but it made the crucial prediction that electric and magnetic fields lived an independent though intertwined existence and could move off into space as a travelling wave at a speed which turned out to identical to that of light. Optics and electromagnetism henceforth had to be considered as one united subject instead of the separate disciplines they had been since the beginnings of Greek science.

Not content with this success, Maxwell proceeded to destroy the machine he had created and build a third model of electromagnetism, a more formal mathematical one which has survived till today. This prodigal feat of the imagination, to disavow two perfectly good, working models for the sake of launching off after new insights is unparalleled in the history of science. The only instance which remotely approaches it is Professor Richard Feynman's battle with quantum mechanics where he has, so far, pioneered three radically different approaches to the subject.

Maxwell became seriously interested in electromagnetism (apart, that is, from decimating the beetle population of Glenlair, p.52) when in his post-Tripos euphoria at Cambridge in 1854, he felt he could return to real science. He wrote to his mentor Thomson:

Trin.Coll., Feb. 20,1854

Dear Thomson,
Now that I have entered the unholy estate of bachelorhood I have begun to think

of reading. This is very pleasant for some time among books of acknowledged merit which one has not read but ought to. But we have a strong tendency to return to Physical Subjects and several of us here wish to attack Electricity.

Suppose a man to have a popular knowledge of electrical show experiments and a little antipathy to Murphy's Electricity, how ought he to proceed in reading and working so as to get a little insight into the subject wh may be of use in further reading?

If he wished to read Ampère Faraday &c how should they be arranged, and at what state & in what order might he read your articles in the Cambridge Journal?

If you have in your mind any answer to the above questions, three of us here would be content to look upon an embodiement of it in writing as advice.[1]

For a philosophically minded physicist, the state of electmagnetism was particularly interesting and the role it played in the development of Maxwell's thoughts has been touched on before (p.63). On the one side were arrayed the mathematicians, trailing in an unbroken line back to Newton (or so they thought) who believed in action at a distance. Over the intervening years they had refined their mathematical techniques and could tackle such problems (in gravity) as the position of a new planet. On the other side of the list was Faraday who knew no mathematics at all, but did not like the 'feel' of action at a distance.

Newton's original pronouncement on this subject, "Hypotheses non fingo" (see p.43) comes from a passage in the *Principia*

> . . . I have not been able to discover the cause of those properties of gravity from phenomena, and I frame no hypotheses . . . it is enough that gravity does really exist, and act according to the laws which we have explained, and abundantly serves for all the motions of the celestial bodies . . .[2]

At that time the idea of action at a distance was revolutionary. Here is a passage from a letter from Huygens to Leibnitz:

> So far as concerns the cause of the tides given by Mr. Newton, I am far from satisfied, nor do I feel happy about any of his other theories built on his principle of attraction, which to me appears absurd.[3]

Descartes had been equally forthright about an earlier theory involving action at a distance, the universal gravitation theory of Roberval:

> The author assumes that a certain property is inherent in each of the parts of the world's matter and that, by the force of this property, the parts are carried toward one another and attract each other. He also assumes that a like property inheres in each part of the earth considered in relation with the other parts of the earth, and that this property does not in any way disturb the preceding one. In order to understand this, we must not only assume that each material particle is animated, and even animated by a large number of diverse souls that do not disturb each other, but also that these souls of material particles are endowed with knowledge of a truly divine sort, so that they may know without any medium what takes place at very great distances and act accordingly.[4]

Yet, as is the way of things, Newton's radical statement had over the years become enshrined in dogma. The process was begun by his disciple, Roger Cotes in the Preface he wrote to the *Principia*, 'action at a distance is one of the primary properties of matter, and . . . no explanation can be more intelligible than this fact'. While he was justifiably displeased at the rush of aetherial fluids to the heads of 18th century natural philosophers, who invoked aetherial mechanisms to account for all the unexplained phenomena of nature, he was bending Newton's own word, as taken here from a letter he wrote to Bentley.

> It is inconceivable that inanimate brute matter should, without the mediation of something else which is not material, operate upon and affect other matter without mutual contact, as it must do if gravitation in the sense of Epicurus be essential and inherent in it . . . That gravity should be innate, inherent, and essential to matter, so that one body can act upon another at a distance, through a vacuum, without the mediation of anything else, by and through which their action and force may be conveyed from one to another, is to me so great an absurdity, that I believe no man, who has in philosophical matters a competent faculty of thinking, can ever fall into it.[5]

Newton went on to feign his own aetherial model of gravity but did not broadcast the idea according to his exegete, Colin Maclaurin, who based this statement on Newton's letters to Robert Boyle:

> It appears, from his letters to Boyle, that this was his opinion early, and if he did not publish it sooner it proceeded from hence only, that he found he was not able, from experiment and observation, to give a satisfactory account of this medium and the manner of its operation in producing the chief phenomena of nature.[5]

Through his extensive knowledge of the history of science, Maxwell was well aware of this. Not only did he quote both the above passages in an article he produced for the Encyclopaedia Britannica, but he later wrote:

> Newton, in his *Principia*, deduces from the observed motions of the heavenly bodies the fact that they attract one another according to a definite law.
>
> This he gives as a result of strict dynamical reasoning, and by it he shows how not only the more conspicuous phenomena, but all the apparent irregularities of the motions of these bodies are the calculable results of this single principle. In his *Principia* he confines himself to the demonstration and development of this great step in the science of the mutual action of bodies. He says nothing about the means by which bodies are made to gravitate towards each other. We know that his mind did not rest at this point–that he felt that gravitation itself must be capable of being explained, and that he even suggested an explanation depending on the action of an etherial medium pervading space. But with that wise moderation which is characteristic of all his investigations, he distinguished such speculations from what he had established by observation and demonstration, and excluded from his Principia all mention of the cause of gravitation, reserving his thoughts on this subject for the "Queries" printed at the end of his "Opticks."
>
> The attempts which have been made since the time of Newton to solve this difficult question are few in number, and have not led to any well-established result.[6]

Maxwell here judiciously ignores Newton's corpuscular theory of light.

Cotes had perverted Newton's intention, 'in spite of his clever exposition of Newton's doctrines, [he] must be considered as one of the earliest heretics bred in the bosom of Newtonianism' as Maxwell put it. Nonetheless, the action at a distance dogma took hold, reaching its high water point in Boscovich's notions. French and German physicists took action at a distance from gravity and applied it to electricity and magnetism; electrostatics is a straightforward inverse square law, although the forces between wires carrying currents and the forces between currents and magnets are more complicated.

Faraday had no formal education and a lifelong fear of mathematical formulae; he could not grasp such complexities and sought for an explanation of electric and magnetic effects in terms he could handle. The beautiful linear patterns formed by iron filings sprinkled on a sheet of paper near a magnet gave him the clue. The filings were affected by the magnet, becoming little magnets themselves and linking end to end. In so doing they traced out lines on the paper, lines of force, Faraday called them, which indicated the direction of the magnetic force on each link of the chain of filings. They traced these lines of force out into space from the magnet and eventually returning to it, the totality of the lines of force giving a 'field' of force.

Instead of one magnet exerting a force on another place some way away, Faraday pictured the first magnet generating these lines of force which leaped out into space twisting, turning and diminishing in strength as they went. The force another magnet experienced then depended on the strength of the lines of force in its locality. Of course, the laws of electrostatic and magnetostatic force stood on firm experimental ground, and the predictions made on action-at-a-distance theory stood untouched—Faraday was really just providing another framework in which to think about the phenomena.

But Faraday's geometrical trellis-work of forces received short shrift from most scientists of the day, such as Sir George Biddell Airy.

> I can hardly imagine anyone who practically and numerically knows the agreement [between Coulomb's laws and experiments] to hesitate one instant between this simple and precise action, on the one hand, and anything so vague and varying as lines of force, on the other hand.[7]

It was undeniable, however, that armed with his 'vague' notions, Faraday was making stupendous experimental discoveries: induction, the electric motor, optical rotation. He had to know something the other physicists did not.

One scientist who took Faraday seriously was William Thomson, ever a champion of ignored causes. In 1841, while an undergraduate at Cambridge, he published a remarkable paper showing that there was a strong formal connection between the mathematical equations which described electrostatics and those which described heat conduction. Since heat conduction certainly involved a continuous flow along lines of temperature gradient, he was, in fact, showing there had to be some substance to Faraday's ideas after all.

It would be natural to suppose that the Maxwell who was to use the standard formulation of action at a distance to such brilliant effect in his work on Saturn's rings (p.110) and gas theory (p.120) would not sit happily in the Faraday camp. Unlike Airy, however, he did recognise Faraday's achievements, and did not wish to prejudge the issue.

> . . . before I began the study of electricity I resolved to read no mathematics on the subject till I had first read through Faraday's Experimental Researches on Electricity. I was aware that there was supposed to be a difference between Faraday's way of conceiving phenomena and that of the mathematicians, so that neither he nor they were satisfied with each other's language. I had also the conviction that this discrepancy did not arise from either party being wrong.

By November 1854, he felt he had absorbed the fundamentals of the subject, as he wrote to Thomson

> Do you remember a long letter you wrote me about electricity, for wh: I forget
> if I thanked you? I soon involved myself in that subject, thinking of every branch
> of it simultaneously, and have been rewarded of late by finding the whole mass
> of confusion beginning to clear up under the influence of a few simple ideas. As I
> wish to study the growth of ideas as well as the calculation of forces, and as I
> suspect from various statements of yours that you must have acquired your
> views by means of certain conceptions which I have found great help, I will set
> down for you the confessions of an electrical freshman.[9]

It is very significant that Maxwell says here that he is interested in 'the growth
of ideas' as well as in electricity, for this was the period in Cambridge when he
was developing his philosophical critique of the inductive method (p.63). As a
referee he was almost always *against* revising papers, to allow the original train
of thought that produced the new idea in the work to shine through. He wrote
as much to Thomson who had announced he was going to publish some of his
electrical works:

> I am glad we are to have some tangible memorial of your experiments for it
> appeared to me that science would suffer for want of a reporter like the worthies
> who lived before Agamemnon.[10]

May 1855 found him working through the action-at-a-distance formulation
of Weber.

> I am reading Weber's Elektordynamische Maasbestimmungen which I have
> heard you speak of. I have been examining his mode of connecting electrostatics
> with electrodynamics, induction &c & I confess I like it not at first . . .
> But I suppose the rest of his views are founded on experiments which are
> trustworthy as well as elaborate.
> I am trying to construct two theories, mathematically identical, in one of
> which the elementary conceptions shall be about fluid particles attracting at a
> distance while in the other nothing (mathematical) is considered but various
> states of polarization tension &c existing at various parts of space. The result
> will resemble your analogy of the steady motion of heat. Have you patented that
> notion with all its applications? for I intend to borrow it for a season.[11]

Maxwell was still hedging his bets, with true philosophic detachment.

By September, however, his own ideas were taking shape, he was choosing
between the different ideas he had read about.

> I have got a good deal out of you on electrical subjects, both directly &
> through the printer & publisher & I have also used other helps, and read
> Faraday's three volumes of researches . . . My object in doing so was of course to
> learn what had been done in electrical science, mathematical & experimental,
> and to try to comprehend the same in a rational manner by the aid of any notions
> I could screw into my head. In searching for these notions I have come upon
> some ready made, which I have appropriated. Of these are Faraday's theory of
> polarity which ascribes that property to every portion of the whole sphere of
> action of the magnetic or electric bodies, also his general notions about "lines of
> force" with the "conducting power" of different media for them.
> Then comes your allegorical representation of the case of electrified bodies by
> means of conductors of heat . . .
> Then Ampère's theory of closed galvanic circuits, then part of your allegory

about incompressible elastic solids & lastly the method of the last demonstration
in your R.S. paper on Magnetism . . .
I do not know the Game-laws & Patent-laws of science. Perhaps the
Association may do something to fix them but I certainly intend to poach among
your electrical images, and as for the hints you have dropped about the "higher"
electricity, I intend to take them.[12]

Thomson was delighted to have so receptive an audience, and Maxwell was
soon able to write to his father 'he is very glad that I should poach on his
electrical preserves'.[13] Maxwell's debt to Faraday was even greater. As he
wrote in the *Treatise:*

> The method which Faraday employed in his researches consisted in a constant
> appeal to experiment as a means of testing the truth of his ideas, and a constant
> cultivation of ideas under the direct influence of experiment . . .
> The method of Ampère, however, though cast into an inductive form, does not
> allow us to trace the formation of the ideas which guided it. We can scarcely
> believe that Ampère really discovered the law of action by means of the
> experiments which he describes. We are led to suspect, what, indeed, he tells us
> himself, that he discovered the law by some process which he has not shewn us,
> and that when he had afterwards built up a perfect demonstration he removed
> all traces of the scaffolding by which he had raised it.
> Faraday, on the other hand, shows us his unsuccessful as well as his successful
> experiments, and his crude ideas as well as his developed ones, and the reader,
> however inferior to him in inductive power, feels sympathy even more than
> admiration, and is tempted to believe that, if he had the opportunity, he too
> would be a discoverer. Every student therefore should read Ampère's research as
> a splendid example of scientific style in the statement of a discovery, but he
> should also study Faraday for the cultivation of a scientific spirit, by means of
> the action and reaction which will take place between the newly discovered facts
> as introduced to him by Faraday and the nascent ideas in his own mind.[14]

No physicist has ever been more open about his methods of research than
Faraday. By studying the various sheafs of Faraday's *Experimental Researches,*
Maxwell came to appreciate the complex way logic, imagination and dogged-
ness intertwine to produce good research. Maxwell expressed this realisation in
his Apostles essay 'Unnecessary Thought' (p.64).

Another general point he learned from Faraday was the importance of the
language in which science is expressed; it should be free of misleading implications and connotations, and should also preferably jolt the reader into new patterns of thought. Maxwell makes the point repeatedly in the *Treatise.*

> In his published researches we find these ideas expressed in language which is
> all the better fitted for a nascent science, because it is somewhat alien from the
> style of physicists who have been accustomed to established mathematical forms of
> thought.[15]

Maxwell repeats the point in Arts. 528 and 567 of the *Treatise.* But most of
all, Maxwell was a convert to Faraday's lines of force:

> If we strew iron filings on paper near a magnet, each filing will be magnetized
> by induction, and the consecutive filings will unite by their opposite poles, so as
> to form fibres, and these fibres will *indicate* the direction of the lines of force.
> The beautiful illustration of the presence of magnetic force afforded by this

experiment, naturally tends to make us think of the lines of force as something real, and as indicating something more than the mere resultant of two forces, whose seat of action is at a distance, and which do not exist there at all until a magnet is placed in that part of the field. We are dissatisfied with the explanation founded on the hypothesis of attractive and repellent forces directed towards the magnetic poles, even though we may have satisfied ourselves that the phenomenon is in strict accordance with that hypothesis, and we cannot help thinking that in every place where we find these lines of force, some physical state or action must exist in sufficient energy to produce the actual phenomena. [16]

Maxwell found that Faraday's vague ideas were in fact capable of formal expression.

As I proceeded with the study of Faraday, I perceived that his method of conceiving the phenomena was also a mathematical one, though not exhibited in the conventional form of mathematical symbols. I also found that these methods were capable of being expressed in the ordinary mathematical forms, and thus compared with those of the professed mathematicians. [17]

Indeed some of the most difficult and powerful techniques that the mathematicians had developed for solving their equations in special cases had a very simple derivation on Faraday's model. Even though it was doubtful that Airy and co. would suddenly admit to seeing the light, the fact that lines of force worked as well as action at a distance was very interesting, philosophically.

The comparison, from a philosophical point of view, of the results of two methods so completely opposed in their first principles must lead to valuable data for the study of the conditions of scientific speculation. [18]

Maxwell's first paper on the subject was read to the Cambridge Philosophical Society in two parts at the end of 1855 and beginning of 1856. The paper began with a longish philosophical introduction (a hallmark of his electromagnetic papers) in which he was careful to apply the lesson he had learned from Faraday's 'alien style' to the use of models and analogies in physics.

To appreciate the requirements of the science, the student must make himself familiar with a considerable body of most intricate mathematics, the mere retention of which in the memory materially interferes with further progress. The first process therefore in the effectual study of the science, must be one of simplification and reduction of the results of previous investigation to a form in which the mind can grasp them. The results of this simplification may take the form of a purely mathematical formula or of a physical hypothesis. In the first case we entirely lose sight of the phenomena to be explained; and though we may trace out the consequences of given laws, we can never obtain more extended views of the connexions of the subject. If, on the other hand, we adopt a physical hypothesis, we see the phenomena only through a medium, and are liable to that blindness to facts and rashness in assumption which a partial explanation encourages. We must therefore discover some method of investigation which allows the mind at every step to lay hold of a clear physical conception, without being committed to any theory founded on the physical science from which that conception is borrowed, so that it is neither drawn aside from the subject in pursuit of analytical subtleties, nor carried beyond the truth by a favourite hypothesis. [19]

Maxwell begins the paper with a new physical model for the lines of force as a straight development of Thomson's mathematical analogy: he connects lines of force with streamlines in fluid flow. An electric charge is equivalent to a pump continuously churning out the fluid. As it flows, the fluid's direction of movement gives the direction of the equivalent lines of force, while the speed of the fluid is equivalent to the intensity of the force. The inverse square law then is just a statement that the fluid is incompressible, and wells out spherically (the area of a sphere goes as the square of the radius). To model the properties of different dielectrics, the fact that the same charge causes different electric fields across different substances, Maxwell assumed that each medium has a different resistance to the flow of the fluid.

What Maxwell did here, therefore, was to develop an extended analogy, finding correspondences between the facts of magnetostatics and electrostatics on the one hand and hydrodynamic flow on the other.

> I do not think that it contains even the shadow of a true physical theory; in fact, its chief merit as a temporary instrument of research is that it does not, even in appearance, *account for* anything.[20]

There was no predictive power to the model, merely a demonstration that there was this one to one correspondence between experiment and the fluid analogy.

What the paper did do, though, was to answer Airy's criticism of vagueness. The laws of electrostatics are equivalent to the laws of hydrodynamics, and the lines of force equivalent to mathematically impeccable stream lines.

> In this outline of Faraday's electrical theories, as they appear from a mathematical point of view, I can do no more than simply state the mathematical methods by which I believe that electrical phenomena can be best comprehended and reduced to calculation, and my aim has been to present the mathematical ideas to the mind in an embodied form, as systems of lines or surfaces, and not as mere symbols, which neither convey the same ideas, nor readily adapt themselves to the phenomena to be explained.[21]

One complicated piece of mathematics did drop out of Maxwell's model in an entirely natural and simple way, and that was what happened at the boundary between two dielectrics, or in the hydrodynamical analogy, between substances of different porosity. He showed the effect was *as if* a surface charge built up at the boundary and wrote down the necessary mathematical boundary conditions which allowed the problem to be solved.

The second part of the paper was quite different. It dealt with the connection between electricity and magnetism; Faraday had discovered that if he moved a loop of wire in the presence of a magnet, a current was induced to flow in it. If there was no magnet present then nothing happened. So Faraday hypothesised that the magnet created a special condition in the wire, the electro-tonic state, so that its subsequent movement was liable to generate currents.

Actually, Faraday later reneged, he found it was not strictly necessary to postulate this special state to explain the observations. But Maxwell would not let him escape so lightly:

> The conjecture of a philosopher so familiar with nature may sometimes be more pregnant with truth than the best established experimental law discovered by empirical inquirers, and though not bound to admit it as a physical truth, we may accept it as a new idea by which our mathematical conceptions may be rendered clearer.[21]

For this electro-tonic state, for the experimental laws of electro-magnetic induction, Maxwell could not yet find a pictorial physical model:

> The idea of the electro-tonic state, however, has not yet presented itself to my mind in such a form that its nature and properties may be clearly explained without reference to mere symbols, and therefore I propose in the following investigation to use symbols freely, and to take for granted the ordinary mathematical operations. By a careful study of the laws of elastic solids and of the motions of viscous fluids, I hope to discover a method of forming a mechanical conception of this electro-tonic state adapted to general reasoning.[22]

He was later to do exactly that, but for the moment he had to revert to a purely mathematical model, and he equated electro-tonicity with a quantity now known as the magnetic vector potential ($\mathbf{A}$).

In itself this was nothing spectacular, Maxwell was merely replacing one variable with another in a sort of mathematical equivalent of what he had achieved pictorially with his hydrodynamic analogy; instead of writing the electromotive force round the loop of wire as being equal to the rate of change of the total magnetic flux through the loop, he could now write it as the rate of change of the vector potential round the loop. What was significant, however, was the form of the mathematics: by using an analytical technique which Thomson had developed (this is what he thanks Thomson for in the letter above, p.138) Maxwell was able to replace the then-current integral equation formulation, by a differential equation. Maxwell knew this was a crucial point.

> We have now obtained in the functions α_0, β_0, $\gamma_0[$ ($\mathbf{A}$)] the means of avoiding the consideration of the quantity of magnetic induction which *passes through* the circuit. Instead of this artificial method we have the natural one of considering the current with reference to quantities existing in the same space with the current itself. To these I give the name of *Electro-tonic functions*.[23]

It was by expressing the laws of electromagnetism in differential form that Maxwell achieved this breakthrough in this subject. When, previously, Ampère had stated that the magnetic force round an imaginary loop with a wire running through its middle is equal to the current carried by that wire, he had used an integral expression (the *total* magnetic force *round* the loop equals the *total* current through it). On three occasions in the paper Maxwell emphasised that Ampère's law had been experimentally verified only in integral form, round complete circuits, 'We know little of the magnetic effects of any currents which are not closed'.[24] It was by going to differential form and considering the ensuing change in Ampère's law, that Maxwell predicted electromagnetic waves.

The local or differential approach to the mathematics of electromagnetism was the natural expression of Faraday's ideas, where lines of force propagate continuously across space, the force in any one element depending on its immediately adjacent neighbours, the effect passing along from element to element along a continuous chain of command. Faraday was uncomfortable with the notion of a force being a summation of the effect of a lot of charges distributed a long way off (Ampère's integral approach), but was interested in the field a charge created in its immediate vicinity, how that field produced a different field slightly farther off, and how that led on: local continuous and contiguous interactions that lend themselves to the mathematics of differential equations.

> In electrical investigations we may use formulae in which the quantities involved are the distances of certain bodies, and the electrifications or currents in those bodies, or we may use formulae which involve other quantities, each of which is continuous through all space.
>
> The mathematical process employed in the first method is integration along lines, over surfaces, and throughout finite spaces, those employed in the second method are partial differential equations and integrations throughout all space.
>
> The method of Faraday seems to be intimately related to the second of these modes of treatment. He never considers bodies as existing with nothing between them but their distance, and acting on one another according to some function of that distance. He considers all space as a field of force, the lines of force being in general curved, and those due to any body extending from it on all sides, their directions being modified by the presence of other bodies.[25]

Maxwell drew a subtle analogy between this and geometry, where a line may be regarded as built up from a number of points or completely equivalently a point may be regarded as the extremity of a line.

These developments were to follow, but for now, the first thing to do, obviously, was to send Faraday a copy of the paper. Faraday was delighted.

> Albermarle Street, W., 25th March 1857.
> MY DEAR SIR—I received your paper, and thank you very much for it. I do not say I venture to thank you for what you have said about "Lines of Force," because I know you have done it for the interests of philosophical truth; but you must suppose it is work grateful to me, and gives me much encouragement to think on. I was at first almost frightened when I saw such mathematical force made to bear upon the subject, and then wondered to see that the subject stood it so well. I send by this post another paper to you; I wonder what you will say to it. I hope however, that bold as the thoughts may be, you may perhaps find reason to bear with them. I hope this summer to make some experiments on the *time* of magnetic action, or rather on the time required for the assumption of the electrotonic state, round a wire carrying a current, that may help the subject on. The time must probably be short as the time of light; but the greatness of the result, if affirmative, makes me not despair. Perhaps I had better have said nothing about it, for I am often long in realising my intentions, and a failing memory is against me.—Ever yours most truly, M. Faraday. [26]

Faraday's *Thoughts on Ray Vibrations* had been delivered eleven years earlier (p.90) and by 1864 Maxwell had read it, for he gives a reference to it in a paper of that year. But the idea of the *time* of magnetic action, which Faraday had finally decided to try to measure, seems to have struck Maxwell like a bolt out of the blue. If electromagnetic effects were not instantaneous that would of course be marvellous ammunition for lines of force, for what could a force be in transit, having left its source but not yet arrived at its target if not some sort of travelling fluctuation along the lines of force? In his enthusiasm for the idea, he used Campbell as a sounding board, poor Campbell having to sit in un-comprehending silence while Maxwell spouted forth his half-developed ideas:

> I wish I could recall the date (1857?) of a drive down the Vale of Orr, during which he described to me for the first time, with extraordinary volubility, the swift, invisible motions by which magnetic and galvanic phenomena were to be accounted for. It was like listening to a fairy-tale.[27]

Regular correspondence between Maxwell and Faraday started up, and Faraday returned Maxwell's compliment by sending him a copy of his next paper, where he wanted to extend the concept of lines of force to gravity as well as electromagnetism. In Faraday's vision, lines of electromagnetic force were under *tension*. A positive charge threw out a line to a negative charge and the tension pulled the two particles together: they attracted each other.

To explain the repulsion between two positive charges, it was necessary to imagine that the lines of force repelled each other. This caused the lines of force to veer away from each other, as in the diagram, most noticeably causing them to run vertically up and down the midway line between the particles.

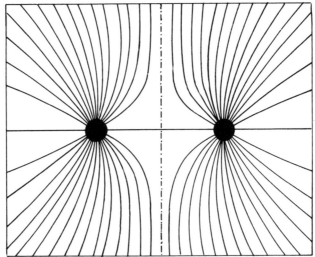

Considering the total tension of the web of lines flung out by just one of the charges, the pull from the deflected lines of force on the side facing the other charge cannot balance the lines of force on the uninterrupted side: there is nothing to pull against line a. There is a net pull away from the side where the lines are deflected which gives the impression that the charges repel each other.

The question Faraday faces was how to extend this analysis to gravity. Maxwell gave the answer immediately.

> . . . you are the first person in whom the idea of bodies acting at a distance by throwing the surrounding medium into a state of constraint has arisen, as a principle to be actually believed in. We have had streams of hooks and eyes flying around magnets, and even pictures of them so beset; but nothing is clearer than your descriptions of all sources of force keeping up a state of energy in all

that surrounds them, which state by its increase or diminution measures the work done by any change in the system. You seem to see the lines of force curving round obstacles and driving plump at conductors, and swerving towards certain directions in crystals, and carrying with them everywhere the same amount of attractive power, spread wider or denser as the lines widen or contract.

You have also seen that the great mystery is, not how like bodies repel and unlike attract, but how like bodies attract (by gravi[ta]tion). But if you can get over that difficulty, either by making gravity the residual of the two electricties or by simply admitting it, then your lines of force can "weave a web across the sky," and lead the stars in their courses without any necessarily immediate connection with the objects of their attraction.

The lines of Force from the Sun spread out from him, and when they come near a planet *curve out from it,* so that every planet diverts a number depending on its mass from their course, and substitutes a system of its own so as to become something like a comet, *if lines of force were visible.*

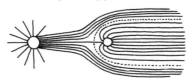

The lines of the planet are separated from those of the Sun by the dotted line. Now conceive every one of these lines (which never interfere but proceed from sun and planet to infinity) to have a *pushing* force instead of a *pulling* one, and then sun and planet will be pushed together with a force which comes out as it ought, proportional to the product of the masses and the inverse square of the distance.[28]

By having all the lines exert a push rather than a pull, gravity could also be included in Faraday's geometric world-view.

Later in the same letter, another passage has a remarkably modern ring to it:

But when we face the great questions about gravitation—Does it require time? Is it polar to the "outside of the universe" or to anything? Has it any reference to electricity? or does it stand on the very foundation of matter, mass or inertia.[29]

The first and last questions stand at the heart of general relativity, while gravity's connection with electricity occupied forty years of Einstein's life, and only recently has any progress been made with the problem (the current answer is 'Yes, possibly').

Faraday's answer was prompt, grateful and totally characteristic:

Albemarle Street,
London, 13th November 1857.

If on a former occasion I seemed to ask you what you thought of my paper, it was very wrong; for I do not think any one should be called upon for the expression of their thoughts before they are prepared, and wish to give them. I have often enough to decline giving an opinion because my mind is not ready to come to a conclusion, or does not wish to be committed to a view that may by further consideration be changed. But having received your last letter, I am exceedingly grateful to you for it, and rejoice that my forgetfulness of having sent the former paper on conservation has brought about such a result.[30]

Faraday recognised clearly the danger of committing oneself to paper too early, with half-formed ideas: once down in writing they act as a nucleus around which one's whole way of thinking crystallises immediately into a solid and unshakeable matrix—which may be the wrong one.

> Your letter is to me the first intercommunication on the subject with one of your mode and habit of thinking. It will do me much good, and I shall read and meditate it again and again . . . I hang on to your words because they are to me weighty . . . There is one thing I would be glad to ask you. When a mathematician engaged in investigating physical actions and results has arrived at his conclusions, may they not be expressed in common language as fully, clearly, and definitely as in mathematical formulae? If so, would it not be a great boon to such as I to express them so?—translating them out of their hieroglyphics.[31]

As Maxwell wrote to Thomson on the following day: 'What a painful amount of modesty he has when he talks about things which may possibly be of a mathematical cast'.[32] Faraday may have been petrified of mathematics but his imagination and intuition were still remarkable. The seed of Time he had planted in Maxwell's brain took root and flowered in his next paper *On Physical Lines of Force* published in three parts over 1861 and 1862.

Here Maxwell took up his self-appointed task, 'to form a mechanical conception of this electro-tonic state'. He now developed a mechanical model of the aether which could explain electromagnetic induction (the connection between moving electric and magnetic fields) as opposed to the simple static properties of the fields, which he had dealt with in his first paper.

His trick was to fill space with microscopic rotating cells whose axes lay along the lines of magnetic force. Maxwell called them 'molecular vortices'. The idea and the name as well came from the work of William Macquorn Rankine, who had in the eighteen-forties produced a model of gases as a sort of fruit jelly: made up of atoms whose nuclei were the pips studding the jelly, each nucleus with an elastic atmosphere spinning around it whose rotational energy constituted the internal energy of the gas. The jelly-like atmospheres all push against each other and against the walls of the vessel enclosing the gas.

It was not a theory which Maxwell found very promising, as he confided to Tait:

> With respect to our knowledge of the condition of energy within a body, both Rankine and Clausius pretend to know something about it. We certainly know how much goes in and comes out and we know whether at entrance or exit it is in the form of heat or work, but what disguise it assumes when it is in the privacy of bodies, or as Torricellis says, "nell' intima corpulenza de' solide naturali", is known only to R. C. and Co.[28]

Yet it was surprising how many conclusions Rankine was able to draw from his model which agreed with experiment. William Thomson, in 1856, took the notion a step further when he supplied a qualitative explanation to one of the most mysterious of Faraday's discoveries on the basis of molecular vortices: in 1845 Faraday had found that if plane-polarised light (light which is, for the sake of argument, vibrating only up and down, only in the vertical *plane*) is shone through a heavy glass block with a strong magnetic field passing through it (in the same direction as the light), then the plane of polarisation of the light is rotated. It might go in vertical, but it will emerge at a slight angle. This was the magneto-optical effect, and a marvellously suggestive finding; how could a magnetic field affect light? Was there a connection between light and magnetism? Perhaps this was one of the strands which led to Faraday's *Thoughts on Ray Vibrations,* but he could make no real progress.

Thomson had far more mathematical ability, and he knew that a straight up and down vibration is mathematically equivalent to the addition of two circular vibrations, one clockwise and the other anticlockwise. If matter was made of molecular vortices aligned by the magnetic field so that they were rotating all in the same sense, then they could interact with the light as it passed through. The vortices would see the plane polarised light as analysed into a mixture of the two circular polarisations, and because they were themselves rotating in one particular sense, let one of these components through more readily than the other (the vortices would act, in a sense, like Polaroid for circular polarisation. Polaroid lets through vertically but not horizontally polarised light). At the far side of the block of glass the two circular polarisations would re-form into plane-polarised light, but they would now be out of phase, one sense of polarisation, having slipped through more easily than the other, would have squeezed in more rotations in its passage through the block. Therefore, when the two components of the light came to recombine, they would no longer meet simultaneously at 12 o'clock, and so their sum would no longer be straight up and down, but would instead be at some angle to the vertical.

Maxwell's model of gases, with lots of freely moving atoms rushing round inside the container, cannoning into the walls, and most of the container filled with *space,* had knocked Rankine's model on the head. But there was a germ of an idea in molecular vortices waiting to be exploited, the association of magnetism with the rotation—of something.

If a body does rotate, then centrifugal force pushes its equator out and gives it a midriff bulge. The earth has just such a bulge: its diameter from pole to pole is *shorter* than a diameter across the equator. Suppose the Universe was filled with an aether made of molecular vortices; small rotating particles which lined up along their poles in a magnetic field. Suppose further that the stronger the magnetic field, the faster they rotate. Then centrifugal force will suck them in along the line of their poles giving a tension, and the equatorial bulge will push them apart along their equators—giving a mutual repulsion between neighbouring lines of force. These are *exactly* the properties needed to explain magnetostatics using the idea of lines of force, as Maxwell had written to Faraday above.

Starting, then, with the idea that there was some magnetic rotation in the 'aether', Maxwell worked out the general mathematical form of the net force on a small volume of the medium and found that it could be broken down quite naturally into a number of contributions.[34] One term gave the natural 'magnetic' inverse square law for a distribution of magnetic 'charges' around the volume. Another term expressed the tendency for paramagnets (diamagnets) to move bodily into (out of) areas of stronger magnetic field. The third term, however, gave a formal contribution to the mechanical force in the x-direction which depended on the magnetic field in the y-direction and also on the vortex velocity (equivalent to the magnetic *force* on Maxwell's hypothesis) taken round the edge of the xy-plane. In the local, differential notation which Maxwell favoured, there was a mechanical force which depended on the curl of the magnetic force in the xy-plane.

Ampère's famous discovery was that just such a circulating magnetic force was generated by an electric current, and so by identifying his 'curl' term with an electric current, Maxwell included Ampère's law in his model, and could then explain the mechanical force on the wire carrying the current that generated the field:

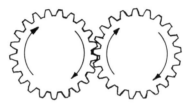

There is a magnetic field from S to N around which the molecular vortices rotate. A current C is pictured as coming straight out of the plane of the paper, and by Ampère's law creates a circulating magnetic field around it. At point E this adds to the magnetic force already present, the molecular vortices spin even faster, their waistlines expand and they push out extra hard laterally. At point W, the two magnetic effects fight each other, the result being a slowing of the vortices and a decrease of lateral pressure. There must therefore be a resultant pressure gradient from E to W pushing the current carrying wire westward. Thus the mechanical force on currents in magnetic fields was explained, and the static magnetic effect of an electric current was united with magnetostatics.

There were still gaping holes in the mechanism of his model: how is the magnetic field geared to the vortex to set it rotating? It was all very well to say that the velocity round the edge of a vortex gave a measure of the electric current through Ampère's law, but how was this to be explained mechanically? Maxwell added another question, the vortices all had to be pictured as rotating in the same sense round the magnetic field, but that should cause friction between the edges of neighbouring vortices.

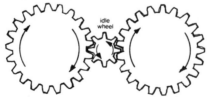

idle
wheel

Engineers faced with such a situation adopt the expedient of the 'idle wheel', they put little counter rotating wheels between the main ones.

Now all the main wheels can rotate, say, clockwise while the idle wheels all go anti-clockwise.

Maxwell proposed to stick idle wheels as lubrication between the main driving wheels of his molecular vortices. Furthermore, rather than pin all the idle wheels to a fixed chassis, he allowed them to move. If, in the diagram, all the vortices are rotating at the same rate, then the idle wheels stay put. But if the angular velocity of the vortices increases as you go to the right, then the idle wheels will all move up the page. The number that move into and out of a small volume depends on how fast the vortices' rotation rate varies across that volume, or, equivalently, how the rotational velocity at the edge of the vortices varies. In fact, it depends on exactly the same curl term that come up in the connection between the electric current and the magnetic force.

> We have, in fact, now come to inquire into the physical connexion of these vortices with electrical currents. . . . Assuming that our explanation of the lines of force by molecular vortices is correct, why does a particular distribution of vortices indicate an electric current?
>
> A satisfactory answer to this question would lead us a long way towards that of a very important one, "What is an electric current?"[35]

Maxwell now took the bold step of identifying the idle wheels with particles of electricity.

Their motion turns the molecular vortices: an electrical current, a flow of the idle wheels along a line, must necessarily cause the idle wheels around that line to rotate, there is a circulating magnetic field around the current. Equally if there is a differential rotation of the vortices, that is, if the magnetic field has a non-zero curl, then the idle wheels move, there is an electric current. Ampère's law was thus now built into the structure of Maxwell's model and was no longer an *ad hoc* addition.

The scene was now set for Maxwell to complete the identification of the laws of electricity with the properties of his model, how to include Faraday's discovery of electromagnetic induction.

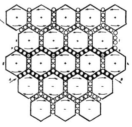

36

Maxwell drew this diagram to illustrate the point; the molecular vortices have a hexagonal cross-section purely for artistic reasons; as he states in the paper, his *calculations* are based on the idea of spherical cells.[37] Next, said Maxwell, let AB be a wire carrying a current. In it the little circular idle wheel particles of electricity are free to move. The wire is surrounded by molecular vortices and more idle wheels.

Now Maxwell considered what happens when the current first starts to move; the vortices in the layer on either side of the wire AB are set spinning, a circulating magnetic force will envelop the wire. But so far it is only the first sheath of vortices which are spinning. All vortices and idle wheels farther out are at rest. As the first set of vortices get going, the idle wheels pq must start to move. They are pinched between a layer of vortices at rest and another moving, there is a differential vortex velocity and so they move bodily. If a conducting wire was placed at pq, which allowed the idle wheels to move freely, then they would now do so, and a current would be induced in the wire. Thus through the intermediary linkage of the molecular vortices, starting an electric current in one wire will induce a current in a parallel wire.

But the wire has resistance, and the current will slow down and come to rest (Maxwell saw non-conductors (insulators) as just having extremely high resistances. There the currents were slowed down so quickly that the idle wheels were permanently trapped within the individual molecules of the insulator, such molecules being many times bigger than the size of a molecular vortex, and so no macroscopic currents would be seen. Maxwell had tried to do an experiment to see if the molecular vortices had a measurable size, but had obtained a null result). If the idle wheels are physically at rest, and the molecular vortices below it still rotating (because the primary current AB is still flowing), then the idle wheels must be themselves spinning and driving the molecular vortices above them to rotate at the same rate. The magnetic field would have been transmitted by the idle wheels to the next layer of vortices, and so the process would continue.

Faraday's law was now explained. Start with a constant magnetic field and stationary, though spinning, idle wheels. Now change the magnetic field in the first layer of vortices. The rim velocities of the vortices in the first and second layers would no longer match and so between them, they would drive the electric charges along, producing a current if there was a wire for it to flow down. But soon that current would stop, and the magnetic field of the second vortex layer would be driven to match the angular speed of the first. So changing magnetic fields produce currents.

Maxwell had now also achieved his ambition of finding a mechanical explanation for the electro-tonic state, which in his first paper he had been obliged to treat only as a formal mathematical proposition: it is the momentum of the idle wheel-electrical charges in an element of the aether.

Having reached this conclusion, Maxwell ended the second part of his paper. With his model he had 'explained' the known experimental laws, and, at a stroke, filled the universe with a complex whirring and clanking aether machine of molecular vortex gears and intermeshed electrical idle wheel cogs.

There is good evidence that he intended to finish the paper here, for he now published it with a recapitulation on what has been achieved. He diligently points out that his model, for all its success, is an *aide-memoir* rather than a belief that Nature is actually constructed this way.

> The conception of a particle having its motion connected with that of a vortex by perfect rolling contact may appear somewhat awkward. I do not bring it forward as a mode of connexion existing in nature, or even as that which I would willingly assent to as an electrical hypothesis. It is, however, a mode of connexion which is

mechanically conceivable, and easily investigated, and it serves to bring out the actual mechanical connexions between the known electromagnetic phenomena; so that I venture to say that any one who understands the provisional and temporary character of this hypothesis, will find himself rather helped than hindered by it in his search after the true interpretation of the phenomena.[38]

Maxwell wrote up his results and then left King's College for his summer vacation at Glenlair.

There, relieved from his routine of lecturing, laboratory supervision and experimental work, away also from all sources of reference, he had the brainwave that led to the prediction of electromagnetic waves. It began with him reconsidering his molecular vortices; they rotate as a whole, therefore they are rigid, yet they can be squashed flatter as they spin faster, therefore they are magnetically elastic too.

Now he reconsidered the electrical idle wheels. Previously he had assumed that they are perfectly free to move around the molecular vortices, but when they come to the boundaries of the gross molecules of tangible matter they experience resistance to their motion, a little resistance for conductors, but a lot for insulators. The resistance was so high in dielectrics, that to all intents and purposes the electric charges stayed permanently put. But from his laboratory experience, Maxwell knew that although there might be no electric currents inside, say, a Leyden jar, the electric field permeates and passes through the dielectric very readily.

> A conducting body may be compared to a porous membrane which opposes more or less resistance to the passage of a fluid, while a dielectric is like an elastic membrane which may be impervious to the fluid, but transmits the pressure of the fluid on one side to that on the other.[39]

The idea was that an electric field caused the idle wheels to move, which caused a mechanical tangential distortion of the vortices. These, being elastic, pulled back against the effect until their force balanced the applied electrical field.

> When the electrical particles are urged in any direction, they will by their tangential action on the elastic substance of the cells, distort each cell, and call into play an equal and opposite force arising from the elasticity of the cells. When the force is removed, the cells will recover their form, and the electricity will return to its former position.[40]

Maxwell now attempted to calculate the electrical elasticity of his medium (E^2 in his notation) and found it depended on how well the medium resisted squashing (m) as opposed to how well it resisted twisting (μ). He did not know its value, but showed it could take on a range of values: for a medium whose elasticity depended only on the forces between the particles. $E^2 = \pi m$, but for a medium with little resistance to shearing (like a jelly) $E^2 = 3\pi m$. So possible values of the electrical elasticity varied by a factor of 3. Maxwell plumped for the lowest value.

> The value of E^2 must lie between these limits. It is probable that the substance of our cells is of the former kind, and that we must use the first value of E^2, which is that belonging to a hypothetically perfect solid. . . .[41]

There seems here to be no trace of philosophic caution about the real existence of the medium, which had tempered the first part of the paper. Maxwell now seems to be a convert to his own medium. And no wonder, for he was now breathlessly leading to his grand result. If the medium is elastic, then if it is given a slight twitch at one point, that twitch will spread out through the medium as a wave, just as plucking a stretched piece of rubber and letting it go will set off vibrations in the rubber. There had to be electromagnetic vibrations in Maxwell's medium.

In any elastic medium, the velocity is the square root of the elasticity divided by the density. Therefore from his calculation of the electrical elasticity of his medium, Maxwell could calculate the speed of these vibrations, and, miraculously it turned out to be the same as the speed of light (the calculated electrical value was 193,088 miles per second, the first experimental value for the speed of light which he locked up was 193,118 miles per second).

> The velocity of transverse undulations in our hypothetical medium, calculated from the electro-magnetic experiments of MM. Kohlrausch and Weber, agrees so exactly with the velocity of light calculated from the optical experiments of M. Fizeau that we can scarcely avoid the inference that *light consists of the transverse undulations of the same medium which is the cause of electric and magnetic phenomena.* [42]

In terms of Maxwell's hexagonal model of the aether, imagine the electrical particles in the wire being jiggled back and forward. The adjacent vortices would be driven clockwise and anticlockwise in turn. Now consider the next layer of electrical particles. Instead of damping out the rotations (because the aether must be an insulator), on Maxwell's new hypothesis, they would be driven bodily to one side, stretching the vortices until their elasticity compensated for the driving force, and then, when this driving force was relaxed, the electrical idle wheels would roll back to their original positions, and then, crucially, overshoot on the other side. And the process *must* then repeat. The physical motion of these electrical particles, as they are displaced from their rest positions, return, overshoot and return again, constitutes a bodily movement of electrical charge in the aether, a current that Maxwell called the displacement current. In the process, the vortices couple this displacement current of charge in one layer of the aether to the charges in the next layer out, which therefore must follow the motion of the charges in the first layer. And they in turn couple to the layer of charge beyond. In this way the initial electrical oscillation in the wire is transferred to a magnetic rotation, to a displacement current in the next layer, to more magnetic rotation, to a further displacement current and so on throughout the medium.

This is an amazing prediction. Here by a rather flimsy *ad hoc* analogy, 'my medium is elastic, all elastic media have waves, therefore my medium must have waves', Maxwell linked together electromagnetism and optics. To make deep predictions after pages of heavy, incontrovertible analysis seems, somehow, a legitimate thing to do. To make earth-shattering predictions purely on the basis of a flimsy 7-line argument (as Maxwell did) requires the confidence, intuition and deep insight of genius. Particularly since his calculations of the elasticity of his medium actually gave an answer that could vary by a factor of 3, and therefore a speed of electromagnetic vibration that could vary by $\sqrt{3}$. And that is apart from having to fill the whole of space with a Meccano set of wheels, axles and cogs.

Nevertheless Maxwell was totally confident in the strength of his result. The rest of that summer holiday must have passed in an enchanted dream and as soon as he returned to London, and could check his calculations, he wrote off euphorically to Faraday.

> I worked out the formulae in the country before seeing Weber's number, which is in millimetres, and I think we have now strong reason to believe, whether my theory is a fact or not, that the luminiferous and electromagnetic medium are one.

and Maxwell concluded the letter with a heartfelt tribute

> When I began to study electricity mathematically I avoided all the old traditions about forces acting at a distance, and after reading your papers as a first step to right thinking, I read the others, interpreting as I went on, but never allowing myself to explain anything by these forces. It is because I put off reading about electricity till I could do it without prejudice that I think I have been able to get hold of some of your ideas . . .[43]

In fact, in the published paper he changed the Fizeau speed of light, which he gave to Faraday as 193,118 miles per second, to 195,647 miles per second—for once not Maxwell's mistake, but that of the author from whom he quoted.

This speed had been measured in air, which as far as light is concerned is almost a vacuum. In more substantial matter, glass, quartz etc., light travels more slowly, and another important result of Maxwell's paper was to relate the speed of light in vacuum to its speed in these materials. This ratio is called the refractive index, and Maxwell showed that it should be equal to the square root of the material's dielectric constant.

Another insight of great importance is also contained in the paper, as Prop. XIV—'To correct the equations of electric currents for the effect due to the elasticity of the medium'. In Maxwell's model the magnetic vortices mesh with the displacement currents, through their gearing to the electrical idle wheels. In Ampère's law the magnetic field was connected to the ordinary laboratory currents in conducting wires.

But Maxwell had already pointed out that Ampère's law was established only for closed circuits. He realised that in differential form, Ampère's law would have to include the displacement current as an extra term on an equal footing with the ordinary current, equally capable of generating magnetic fields. As an added bonus, he pointed out, the electrical continuity equation—the total electric current out of any given volume must equal the rate of change of the electrical charge in that volume— would also now be obeyed, which it could not be on the previous formulation of Ampère's law, a fact which everyone seems to have overlooked.

Part IV, the last part of the paper, considers Faraday's magneto-optical effect, completing the circle by analysing the effect which had originally led him to consider molecular vortices. The formula he worked out is complicated, but highly significant, in that it predicted that the effect depends linearly on the radius of the molecular vortices. This is the *only* place in the whole paper where the properties of the vortices may actually directly play a part in some experimental phenomenon; everywhere else the vortices were used to explain what was going on, but dropped out of the final answers, leaving them solely dependent on macroscopic, directly measurable quantities.

It was here that Maxwell finished what surely must rank as one of the greatest papers ever written. An enormous step in the development of physics had been made, though, inevitably, there were still many loose ends to be tied off. Firstly, the predicted speed of electromagnetic vibrations had to be checked. This involved the comparison of the electrostatic unit of current with the electromagnetic, and was one of the problems he tackled under the aegis of the B.A.

At a much deeper level, however, Maxwell had to consider the status of his model of the aether. It had led him to make a number of predictions but nowhere (except for the somewhat peripheral magneto-optical effect) did it enter those predictions in a crucial way—unlike the molecular theory of gases where explanations for viscosity, diffusion, thermal conduction, effusion etc. etc. had immediately followed. Most scientists would nonetheless have rested on their laurels: having achieved such marvellous results on the basis of a model, it would have been unthinkable that that model could be wrong in any but minor details, they would have been quite incapable of going back and rethinking the whole basis of the model. Maxwell was well aware of the danger of an apparently successful theory, as he showed in a discussion of the contributions Rankine had made to thermodynamics on the basis of *his* molecular vortex theory

> [the scientist] must imagine model after model of hypothetical apparatus till he finds one which will do the required work. If this apparatus should afterwards be found capable of accounting for many of the known phenomena, and not demonstrably inconsistent with any of them, he is strongly tempted to conclude that his hypothesis is a fact, at least until an equally good rival hypothesis has been invented. Thus Rankine, long after an explanation of the properties of gases had been founded on the theory of the collisions of molecules, published what he supposed to be proof that the phenomena of heat were invariably due to steady closed streams of continuous fluid matter.[44]

But this is exactly where Maxwell's philosophical caution came through, (despite its taking an occasional cat nap (p.151)). It was up to experiment to check the validity of the theory, and the theoretician to find incontrovertible experimental predictions for the experimenter to check.

> If, by the same hypothesis, we can connect the phenomena of magnetic attraction with electromagnetic phenomena and with those of induced currents, we shall have found a theory which, if not true, can only be proved to be erroneous by experiments which will greatly enlarge our knowledge of this part of physics.[45]

Maxwell now set about reworking his results in electromagnetism to see where, if at all, molecular vortices were crucial.

He had precedents here. In thermodynamics, the 2nd law (or its equivalent), which was undoubtedly correct, had been derived at least twice on the basis of models that were undoubtedly incorrect: Rankine had done it for molecular vortices, and earlier the Frenchman Sadi Carnot had discovered the law on the basis of the Caloric fluid theory of heat. So Maxwell was well aware of the remarkable ability of science to 'bootstrap' itself off the ground: had it happened here?

It took him three years before he had his answer—yes. Towards the end of 1864 he was able to write to Kelvin

> I can find the velocity of transmission of electromagnetic disturbances independently of any hypothesis now and it is $= v$[(the speed of light)] and the disturbances must be transverse to the direction of propagation or there is no propagation thereof.[46]

He had no need of his aether model now to understand the speed of light and no uncertainties about its exact value either. For once he could not restrain his delight

> I have also a paper afloat, with an electromagnetic theory of light, which, till I am convinced to the contrary, I hold to be great guns.[47]

He was not exaggerating, it is a great paper.

In his 1861 paper, he had taken the experimental laws of electromagnetism and interpreted them in terms of a model. He had then gone a step further: his model gave him elastic vortices which should be able to wobble, therefore there should be electromagnetic vibrations.

Now, in *A Dynamical Theory of the Electromagnetic Field* which he read before the Royal Society in December 1864,[48] he went from the differential laws of electromagnetism, with only the electric and magnetic fields which occur there as his variables, by a direct mathematical derivation to a wave equation: no worries about whether the medium was an ideal solid, no mention of the medium at all:

> I have on a former occasion attempted to describe a particular kind of motion and a particular kind of strain, so arranged as to account for the phenomena. In the present paper I avoid any hypothesis of this kind; and in using such words as electric momentum and electric elasticity in reference to the known phenomena of the induction of currents and the polarization of dielectrics, I wish merely to direct the mind of the reader to mechanical phenomena which will assist him in understanding the electrical ones. All such phrases in the present paper are to be considered as illustrative, not as explanatory.[49]

Maxwell himself later compared these two electromagnetic papers thus

> The former is built up to show that the phenomena are such as can be explained by mechanism. The nature of the mechanism is to the true mechanism what an orrery is to the Solar System. The latter is built on Lagrange's Dynamical Equations and is not wise about vortices.[50]

The orrery models the motions of the planets, and, if well constructed could presumably be used to give accurate predictions of their movements. But the clockwork mechanism of an orrery, such as the mechanical model Maxwell did actually build for the rings of Saturn, bears no resemblance at all to the mechanism of gravity: the orrery is a help in visualising planetary motion, not an explanation for it.

What Maxwell had done in his new paper he compared to being a bell-ringer

> In an ordinary belfry, each bell has one rope which comes down through a hole in the floor to the bellringer's room. But suppose that each rope, instead of acting on one bell, contributes to the motion of many pieces of machinery, and that the motion of each piece is determined not by the motion of one rope alone,

but by that of several, and suppose, further, that all this machinery is silent and utterly unknown to the men at the ropes, who can only see as far as the holes in the floor above them.[51]

All the bell ringer can do is find out which bells peal when particular ropes are pulled: which final effects are connected to particular initial causes, while the detailed gearing of the bell-tower could remain forever unknown.

In his 1861 paper Maxwell made one crucial macroscopic change to the experimental input laws of electromagnetism that was not just an arbitrary detail of his model, he had added the displacement current to the ordinary current in Ampère's law. It was vital to check whether this extra term was justified. As early as December 1861, having just completed his mechanical model, he was checking to see whether there could be any other possible variation in Ampère's law.

> I am trying to form an exact mathematical expression for all that is known about electro-magnetism without the aid of hypothesis, and also what variations of Ampère's formula are possible, without contradicting his expressions . . . So I am investigating the most general hypothesis about the mutual action of elements, which fulfils the condition that the action between an element and a closed circuit is null.[52]

He convinced himself that his contribution to the law was correct, and was the only extra term needed in going from the integral form of the laws to the differential form he favoured. And thus for the first time, in the first sections of his 1864 paper he wrote down the block of general differential equations connecting the components of the electromagnetic field, which today have become known as Maxwell's equations.

It was just this simple extra term, the displacement current in Ampère's law, which meant that when the four Maxwell equations were combined, electromagnetic waves should exist. Maxwell looked at this question in part VI of the paper. There he demonstrated very clearly that when these equations were combined, a wave solution should exist, that its speed should be the speed of light (or at least the esu/emu ratio) and that these electromagnetic waves could be only transverse. There could be no doubt now.

This had been a big stumbling block for most mechanical attempts at a theory of light (of which there had been many and were to be many more in Victorian times): with all known materials, compressional waves (i.e. sound waves) are possible. Only if the material is sufficiently rigid 'sideways', could 'shakes' also propagate as waves—as transverse waves. For 'soft' materials, such as water, which have no resistance to shearing forces, no transverse waves are possible, but compressional waves can still travel very well. The aether was unique in that it carried transverse waves, but, seemingly, no 'sound', and to get around this problem, mechanical aether theories had to perform the most extreme and unnatural contortions. Transverse electromagnetic waves just dropped out of Maxwell's theory with no trouble at all. (In fact, Maxwell realized there was an extra degree of freedom in his equations which enabled him to transform his variable into a particularly neat form—he made what is today known as a gauge transformation to the Coulomb gauge—in this paper).

As a demonstration of the power of his equations, Maxwell proceeded to the theory of double refraction, the phenomenon where certain crystals split an incident beam of light into two components polarized at right angles to each other. It was by the cunning application of this effect that Nicol had invented the Nicol prism to use as a polarizer of light. The French theoretician Augustin Fresnel had *guessed* a solution to the way double refraction works earlier in the century, but Maxwell was now able to *calculate* the answer unambiguously from his theory. He confirmed the formula from his earlier paper for the refractive index of a medium, and also showed that light carries momentum as well as energy, so that sunlight reflecting off a mirror should push against it,

> Such rays falling on a thin metallic disk, suspended delicately in a vacuum, might perhaps produce an observable mechanical effect.[52]

he wrote in the Treatise, but the effect should be tiny: a mere 4lbs weight for a square mile of disk. No wonder he was stunned when he witnessed Crookes' radiometer.

Underlying his hypothesis-free mathematics *was* a firm belief in *an* aether

> We may therefore receive, as a datum derived from a branch of science independent of that with which we have to deal, the existence of a pervading medium, of small but real density . . .[54]

which transmitted the light. But he was careful to keep facts and hypotheses firmly separate. The only term he used in the paper to which he was prepared to commit himself unquestioningly was Energy, as quoted above all others were illustrative not explanatory, but still

> The only question is, Where does it [(energy)] reside? . . . On our theory it resides in the electromagnetic field, in the space surrounding the electrified and magnetic bodies as well as in those bodies themselves, and is in two different forms, which may be described without hypothesis as magnetic polarization and electric polarization, or, according to a very probable hypothesis, as the motion and strain of one and the same medium. The conclusions arrived at in the present paper are independent of this hypothesis.[55]

The medium could still, presumably, make its presence felt in the magneto-optical effect, about which Maxwell says nothing in this paper, and also in the compressional vibrations of the medium. His equations say that electromagnetic vibrations must be transverse; the ability of the medium to transmit 'sound' waves is open

> Electromagnetic science leads to exactly the same conclusions as optical science with respect to the direction of the disturbances which can be propagated through the field; both affirm the propagation of transverse vibrations, and both give the same velocity of propagation. On the other hand, both sciences are at a loss when called on to affirm or deny the existence of normal vibrations.[56]

One other avenue which Maxwell found closed to him was gravity—about which he had originally written encouragingly to Faraday. The trouble was that gravity is attractive. The energy density of the field was the one sacrosanct concept for Maxwell, and it is given by $C-R^2$ where R is the gravitational force and C is an arbitrary constant. Maxwell felt that energy had to be 'essentially positive, it is impossible for any part of space to have negative intrinsic energy'. Hence C had to be adjusted so $C-R^2$ had its lowest value of O where R was the strongest gravitational force in the Universe. Unfortunately with C chosen thus, the energy density of the rest of the universe would have to be absolutely enormous, and Maxwell could not conceive how any medium could be so constructed.

If Maxwell found difficulties only with infinite gravitational energy densities, his contemporaries had quite enough to chew on with his paper. Just as with the velocity distribution of gases, Maxwell was waving his magical wand and producing magical results. In his earlier paper, he had produced a mechanical model of the aether, and the other physicists were happy enough with that, this was a model with which their Victorian imaginations could come to grips. Indeed plenty of them had themselves made attempts to construct just such models. The 1864 paper cheated; it got all the results for which they were searching so hard with no mention of anything but the electric and magnetic fields, nothing at all about the aether, almost as if the aether which they felt to be one of the surest and most certain things in the Universe were not there at all

> If I knew what the electromagnetic theory of light is, I might be able to think of it in relation to the fundamental principles of the wave theory of light. But it seems to me that it is a rather backward step from an absolutely definite mechanical notion that is put before us by Fresnel and his followers to take up the so-called Electro-magnetic theory of light in the way it has been taken up by several writers of late . . . I merely say this in passing, as perhaps some apology is necessary for my insisting upon the plain matter-of-fact dynamics and the true elastic solid as giving what seems to me the only tenable foundation for the wave theory of light in the present state of our knowledge.[57]

said William Thomson! Maxwell's idea of looking at the field equations was very slow to break through the prejudices of fellow physicists.

Maxwell's next paper (1868) was partly a more easily understood derivation of his main result, with the complication of the gauge transformation omitted, and partly an attack on the action-at-a-distance school. The German physicists had derived a formulation of electromagnetism where the long-range forces depended on the velocity and acceleration of the charges as well as their distances. Maxwell considered two particles, a fixed distance apart, moving together along their axial line. The particle in front moves away from where its partner was, that is to where its partner's force was weaker. The one behind is moving always into a region of stronger force. The forces therefore cannot match, and one has a breakdown of the laws of conservation of energy unless one also allows the electric field to carry momentum and energy

> I think that these remarkable deductions from the latest developments of Weber and Neumann's theory can only be avoided by recognising the action of a medium in electrical phenomena.[58]

Maxwell's splendid rudeness here is matchèd only by the courtesy with which he points out an experimental weakness of Weber's method of measuring the esu/emu ratio, i.e. the speed of light

> I mean to indicate that I have such confidence in the ability and fidelity with
> which their investigation was conducted, that I am obliged to attribute the
> difference of their result from mine to a phenomenon the nature of which is now
> much better understood than when their experiments were made.[59]

Maxwell's own value for the ratio, 288,000,000 metres per second is almost
exactly as innaccurate as Weber's was.

Since his ideas on electromagnetism were not getting the audience they
deserved, Maxwell decided that something more substantial in the way of
propaganda was needed. It was partly with this in mind that he decided to
write the *Treatise on Electricity and Magnetism* (1873), the massive textbook
that he spent so much of his time at Glenlair in compiling.

It is a magnificent book, many sections of which could still be used today
(some parts of it are better than today's texts). It reworks in comprehensive
detail just about all that was known in electricity and magnetism, experiment,
theory and instrumentation, leading up to Maxwell's own synthesis of this
knowledge to produce his electromagnetic unification.

En route, as a by-blow, Maxwell develops the theory of spherical harmonics,
the mathematics of conjugate functions, and introduces vector notation (in
minimal quaternion form)*

It seems quite remarkable that Maxwell's book which even today is a model
of precision and clarity should have appeared murky and impenetrable to his
contemporaries. Henri Poincaré the great French mathematician said 'I
understand everything in the book except what is meant by a body charged
with electricity'. Henrich Hertz, the equally great German physicist also had
his problems

> Many a man has thrown himself with zeal into the study of Maxwell's work, and,
> even when he has not stumbled upon unwonted mathematical difficulties, has never-
> theless been compelled to abandon the hope of forming for himself an altogether
> consistent conception of Maxwell's ideas. I have fared no better myself.[62]

Hertz continued this comment with 'What is Maxwell's theory? I cannot give
any clearer or briefer answer than the following: Maxwell's theory is the
system of Maxwell's equations'.

* In so doing he ran slap into the bugbear of all students of vector quantities: a right-handed
screw is not a left-handed one
>   O T' I am desolated! I am like the Ninevites! Which is my right hand? Am I perverted? a
>   mere man in a mirror, walking in a rain show? What saith the Master of Quaternions? i to
>   the South, j to the West and k to the Heavens above. See lectures §65 kT . . . But what
>   saith T and T' §234 They are perverted. If a man at Dublin finds a watch, he lays it on the
>   ground with its face up and its hands go round from S to W and he says this is a + rotation
>   about an axis looking upwards. If the watch goes to Edinburgh or Glasgow T' or T
>   carefully lays it down on its face and after observing the gold case he utters the remarkable
>   aid to memory contained in §234 of the book[60]
T and T' in their book were obviously somewhat unclear. Maxwell then carried out some
fundamental research into handedness which ended up as a footnote in the Treatise 'the tendrils of
the vine are right-handed screws, and those of the hop left handed' (a fact which Flanders and
Swan have used most poetically) 'the system of the vine, which we adopt, is that of Linnaeus, and
of screw makers in all civilised countries except Japan . . . Screws like the hop tendril are made . . .
for the fittings of wheels on the left side of ordinary [(railway)] carriages'[61]

Here, I think, lies the clue to the mystery. The *Treatise* is written in Maxwell's best positivist style, as is his 1865 paper: he tries to stick as much as possible to experimentally measurable quantities; his theory *is* just his equations and by and large, Victorian scientists were just not ready for this. Maxwell's personal belief in an aether is not allowed to prejudice the book. His own model makes only the briefest of appearances

> The attempt which I . . . made to imagine a working model of this mechanism must be taken for no more than it really is, a demonstration that mechanism may be imagined capable of producing a connexion mechanically equivalent to the actual connexion of the parts of the electromagnetic field. The problem of determining the mechanism required to establish a given species of connexion between the motions of the parts of a system always admits of an infinite number of solutions.[63]

The nature of electricity was also hypothetical: the one-fluid or two-fluid theories (i.e. that there was just one sort of 'fluid' electricity, so that when a body is + charged it has acquired too much of the fluid, when it is – charged it has lost too much fluid, or that there are two sorts of 'fluid', and whichever charge a body has acquired merely indicates which fluid it has been drinking up too much of). Maxwell introduced both theories but took no sides

> In the present treatise I propose, at different stages of the investigation, to test the different theories in the light of additional classes of phenomena. For my own part, I look for additional light on the nature of electricity from a study of what takes place in the space intervening between electrified bodies.[64*]

Maxwell actually did not believe in either theory, he followed Faraday's notions, the 'no-fluid theory' that charge was a snare and a delusion. He believed that when an electromotive force was applied to any dielectric (including free space) there followed a *displacement* of electricity in that dielectric. What was observed as a charge was merely a surface effect at the boundaries between dielectrics, where the different displacement did not match. He wrote to Thomson:

> It follows from this theory that the movements of electricity are like those of an incompressible fluid, and that charging a body and putting it into a cubic foot of air does not increase the quantity of electricity in the cubic foot.[67]

But he did not want to proselytize even his own beliefs; he wanted the true theory of electricity to grow incontrovertibly out of experiment. Perhaps Poincaré's statement should be taken as a compliment.

---

* In his later work, *An Elementary Treatise on Electricity*, Maxwell not only issues a far sterner warning aginst the idea of electric fluids

> here we may introduce once for all the common phrase *The Electric Fluid* for the purpose of warning our readers against it. It is one of those phrases, which, having been at one time used to denote an observed fact, was immediately taken up by the public to connote a whole system of imaginary knowledge . . . we must avoid speaking of the electric fluid.[65]

and also reports on some experiments he had done on the conduction of electricity in gases. He let the vapour of boiling mercury and sodium pass between a pair of electrodes, but took care that no gas from the flame of the Bunsen burner used to heat them did so too. The electrodes did not discharge, whereas exposed to the Bunsen burner gases they did so immediately. This he found mystifying.[66]

The one hypothesis which he found inescapable was that something moving is to be associated with electrical currents.

> The electric current cannot be conceived except as a kinetic phenomenon. Even Faraday, who constantly endeavoured to emancipate his mind from the influence of those suggestions which the words 'electric current' and 'electric fluid' are too apt to carry with them, speaks of the electric current as 'something progressive, and not a mere arrangement.'[68]

It is with this clue that the Treatise suddenly changes gear half-way through the second volume. Instead of a magisterial unravelling of experimental fact, it becomes a remarkable derivation of the existence of electromagnetic waves from the simple recognition that something dynamical is going on.

> It is difficult, however, for the mind which has once recognised the analogy between the phenomena of self-induction and those of the motion of material bodies, to abandon altogether the help of this analogy, or to admit that it is entirely superficial and misleading. The fundamental dynamical idea of matter, as capable by its motion of becoming the recipient of momentum and of energy, is so interwoven with our forms of thought that, whenever we catch a glimpse of it in any part of nature, we feel that a path is before us leading, sooner or later, to the complete understanding of the subject.[69]

Now Maxwell reworked the facts of electromagnetism painstakingly built up in the first one and a half volumes of the Treatise from the same dynamical viewpoint as his 1864 paper. The significance here was that it enabled him to use the generalised Lagrangian formulation of dynamics.

Lagrange had developed the technique to deal with mechanical systems, but Maxwell recognised that it was of much more general application, and he could apply it to the time variation of his electromagnetic fields. Its attraction was that very few assumptions need be made, a general mathematical function was postulated—on symmetry grounds, or just to include everything one could think of—and once that was done, a routine mathematical program dropped out the results. Maxwell's bell-ringer analogy (p.155) was actually a comment on the Lagrangian method. It was the perfect mathematical tool for his positivist needs.

He began with full generality, including electromagnetic and ordinary mechanical terms in his Lagrangian. He realised that there might also be crossed terms, where his electromagnetic fields might get mixed up with ordinary mechanical terms and the elctromagnetic fields might interfere with the *motion* of magnets and conductors in their vicinity. He identified three possible effects, one of which he had himself tried to detect experimentally back in 1859, long before this formal analysis was carried out. The idea was that if a magnet contained aligned magnetic vortices, then these should also contribute to its angular momentum. If, when spinning fast, its stability could be made precarious, this extra angular momentum could just tilt the balance between the alternate positions of stability. In this way he had hoped to get some measure of the reality of molecular vortices, but his apparatus had been too insensitive (all three of his postulated effects were later detected). On his new analysis, the test showed merely a coupling between electromagnetic and mechanical variables, and so, if it had been detected, would have actually revealed nothing about molecular vortices; by his Lagrangian approach, he showed that the vortices were not strictly necessary to have the coupling.

The same thing happened in his analysis of the magneto-optical effect, which had been detected experimentally and which he had shown to depend critically on the radius of the molecular vortices in his 1861 paper. In the Treatise he destroyed his own case, with another application of Lagrangian analysis, based on the symmetry of that function: it allowed only certain sorts of mathematical expression to appear. From this minimal hypothesis he deduced a formula for the angle of rotation in the magneto-optical effect which depended only on macroscopic, measurable terms; there was no place in it for the vortex radius. But, he added in a note, he still felt that Thomson's argument that something had to be rotating in the glass block was valid—it was just that its precise nature did not appear in the final formula. (Today one would say that the atomic electrons were rotating about the magnetic field. Maxwell's style of analysis here, based on Lagrangians and general symmetry considerations, is one that theoretical physicists have made much use of in the récherché recesses of elementary particle physics in the 20th century).

Clearly the Treatise was not merely a standardization of well-known ideas, it was a restatement of them from his own special viewpoint, and as with the magneto-optical effect, a great development of them. There is a Cambridge story that Maxwell said he wrote the Treatise 'to educate himself by the presentation of a view of the stage he had reached'[70] which is probably far nearer the truth than his own modest statement that his book was a mere (913 page) mathematicisation of Faraday's ideas.

Apart from the remarkable coincidence of the speed of light with the esu/emu ratio, Maxwell's theory had little to support it when the Treatise was first published. The best Maxwell could offer was some tentative evidence that for melted paraffin, the square root of the dielectric constant (ε) was not far away from its refractive index (μ). The problem was that the refractive index depends on the frequency of light (which is why prisms bend the differently coloured constituents of white light by different amounts to produce the rainbow). Measurements of the dielectric constant were made with static fields, i.e. with a frequency of zero. To compare these two experimental parameters, therefore, the refractive index had to be extrapolated from the frequency of light, about 10^{15} vibrations per second, to zero frequency. There was just no way of knowing how that extrapolation should be carried out, and the fact that the values agreed even approximately ($\sqrt{\varepsilon} = 1.405$, $\mu = 1.422$) was something of a miracle.

More miracles soon followed. One of the few physicists who took Maxwell's ideas seriously was Helmholtz in Berlin. He set his young student Ludwig Boltzmann (no less) to check Maxwell's formula. The experimental results turned out to be all over the place, and they thought they had disproved Maxwell's theory. Then Boltzmann looked at his results again and realized he had been trying to prove the wrong thing. He and Helmholtz together had misread Maxwell and thought his theory predicted that the refractive index should be directly proportional to the dielectric constant

> On now putting the values of all the dielectric constants together in a table, I
> was much worried at their deviating so far from the refractive indices, but
> noticed at the same time that they were always nearly equal to the square of the
> latter. The thought flashed through my mind that Maxwell's theory might
> require this, since the velocities of transmission are always proportional to the
> square root of the forces. I looked up Maxwell's treatise, and there sure enough
> was plain to read that the dielectric constants must be proportional to the
> squares of the refractive indices.[71]

Maxwell and his students did similar experiments themselves at Cambridge.

The same problem of frequency extrapolation stymied another test of the theory: the amount of light transmitted by gold leaf. Maxwell calculated that only 10^{-300} of the incident light should get through, an invisible small amount, yet a greenish glow is clearly visible through a thin piece of leaf.[72] Maxwell was using the value for the static electrical conductivity of gold.

One clear piece of evidence for the electromagnetic theory of light was missed by Maxwell: the formula for the ratio of the amounts of light reflected and refracted when an incident beam strikes the interface between two transparent substances. Fresnel had guessed the solution to the problem, but it is directly calculable on the electromagnetic theory, and is no more than a boundary problem of the sort Maxwell had solved in his 1855 paper.

In fact when he had submitted his paper of 1864 to the Royal Society, Stokes had suggested that this was a problem he could usefully investigate. Maxwell first replied that he was not 'inclined' to do it, preferring to accept the experimental data. He then must have had second thoughts, because in his next letter, he says at the beginning that he is not sure what are the correct boundary conditions to impose, and at the end of the letter brings up the frequency problem again

> I am trying to understand the conditions at a surface for reflexion and refrac-
> tion, but they may not be the same for the period of vibration of light and for
> experiments made at leisure.[73]

Maxwell did have a go at the calculation, but did not publish it, and it was left for the Dutch physicist Hendrik Lorentz (1874) to solve the problem; he simply imposed the obvious boundary conditions from Maxwell's equations, and showed the theory predicted exactly Fresnel's results.

Underlying the whole theory was the basic ambiguity of the nature of electric charge; this was on the verge of being solved when Maxwell died. In the Treatise Maxwell had predicted that when the passage of electric currents through tenuous gases 'are better understood they will probably throw great light on the nature of electricity' (Art. 57), and it was just such experiments that led to the new understanding. As a referee, Maxwell had watched the development of Crookes' experimental techniques for studying the radiometer effect, correctly pointing out that the achievement of better and better vacua was the most important scientific gain our of the whole procedure. Maxwell was also the referee when Crookes went on to use his vacua to study gas discharges.

What Crookes observed is now interpreted as an electrical breakdown in the gas, a high voltage applied across a near vacuum ripping electrons off the gas molecules, these electrons then being accelerated towards the cathode, almost unimpeded, because of the tenuousness of the gas, and reaching extremely high speeds because of their light mass. But Maxwell could not know then of the existence of the electron. This is his report

> That a powerful action is going on is proved . . . by the far more remarkable fact that the emanation from the negative electrode when directed on a piece of platinum was able to raise it to a white heat. If this was done by the impact of molecules projected from the negative electrode the velocity of these molecules must have been far greater than that of the molecules of a gas at white heat, say 1500°C . . . Mr. Crookes has shown 1st that the emanation is deflected when it passes transversely across lines of magnetic force as if a transverse force acted on a moving body similar to that which acts on a conductor carrying an electric current when in the same field of force.
>
> Now Professor H. Rowland has shown that electrified bodies in motion produce the same effect on a magnet as an electric current in a conductor, so that we might suppose the emanation to consist of a stream of molecules every one of which has a negative charge.[74]

Maxwell went on to suggest to Crookes the experiment later performed by Perrin in France of shooting the 'emanation' into an electrometer and measuring the charge. He also worked out that the fact that the magnetic deflection of the beam was large while its electrostatic deflection was immeasurable, was exceedingly odd—the obvious implication being that the emanation is travelling at an absolutely enormous velocity.

Some of Maxwell's other comments have a typical ring:

> . . . and as Faraday entitled one of his greatest papers 'On the Magnetization of Light and the Illumination of Magnetic Lines of Force" so Mr. Crookes calls his discovery "An Illumination of the Lines of Molecular Pressure". Whether either of these authors chose the best form of words for enlightening their contemporaries may be questioned, but when so much that we can appreciate is set before us we may accept and even preserve as a souvenir a metaphor struck off in the heat of discovery. . . .
>
> It is probably impossible, in describing, day by day, such a course of investigations, to avoid the use of expressions involving hypotheses. The hypotheses, however, which Mr. Crookes has used in devising his experiments are distinct enough to have guided him to the discovery of new phenomena and yet they do not appear, either in the author's writings or in what they reveal of his mind, to have hardened into dogmas from which he is not prepared to withdraw when reason bids him.[74]

Maxwell was on the verge of the discovery of the electron. If he had not been so faithful to Faraday's ideas, he might well have taken his own intuition more seriously; his aether was full of small electrically charged particles.

He came even closer to the electron in his chapter on electrolysis in the Treatise.

Electrolysis is the decomposition of conducting substances by the passage of an electric current. Humphry Davy who was Faraday's boss at the Royal Institution had discovered the metal sodium by electrolysing sodium chloride:

sodium (the cation) collects at the positive electrode (the cathode). Together, Davy and Faraday had later discovered the element iodine also by electrolysis, using their portable physical laboratory while jaunting around France oblivious of the fact that Napoleon was at war with Britain at the time. Faraday had gone on to perform his own electrolytic experiments, and deduced his great law that the chemical equivalent weight of any substance is always released by the same quantity of electricity, which would now be interpreted as the same number of atoms of any element (albeit divided by their valency) being released by the same quantity of electricity.

To the scientists of the day this seemed a magical result, and it waited until the publication of the Treatise for its 'natural' interpretation

> It is therefore extremely natural to suppose that the currents of the ions are convection currents of electricity, and, in particular, that every molecule of the cation is charged with a certain fixed quantity of positive electricity, which is the same for the molecules of all cations, and that every molecule of the anion is charged with an equal quantity of negative electricity . . . we call this constant molecular charge, for convenience in description, *one molecule of electricity*. This phrase, gross as it is, and out of harmony with the rest of this treatise, will enable us at least to state clearly what is known about electrolysis, and to appreciate the outstanding difficulties.[75]

Maxwell was prepared to use the idea as an unifying model of electrolysis, but was too cautious to take it seriously. It was not until 1881 when Helmholtz gave a lecture in London to the Chemical Society, that someone took the 'molecule of electricity' as a fact and actually pictured chemical compounds as being made of electrically charged atoms bound together by their charges. Maxwell had in fact already considered this idea, and rejected it as being too facile:

> The fact that every chemical compound is not an electrolyte shows that chemical composition is a process of a higher order of complexity than any purely electrical phenomenon.[76]

This chapter will close with one more example of Maxwell's catlike intuitive caution. It comes from his very first paper on electromagnetism in 1855.

> The changes of direction which light undergoes in passing from one medium to another, are identical with the deviations of the path of a particle in moving through a narrow space in which intense forces act. This analogy, which extends only to the direction, and not to the velocity of motion, was long believed to be the true explanation of the refraction of light; and we still find it useful in the solution of certain problems, in which we employ it without danger, as an artificial method. The other analogy, between light and the vibrations of an elastic medium, extends much father, but, though its importance and fruitfulness cannot be over-estimated, we must recollect that it is founded only on a resemblance *in form* between the laws of light and those of vibrations.[77]

The analogy of light to a wave, which he himself developed to a new mathematical level, is still only an analogy. Maxwell was already mentally geared for the discovery of quantum effects in optics.

CHAPTER 12

Maxwell's *other* contributions to physics stand inevitably in the shadow of the twin peaks of electromagnetism and molecular gas theory. Yet many of them are major advances in their own right, and they shed an interesting light on his genius.

His first publication was on geometry—the paper on ovals he wrote as a schoolboy, and all of his early theoretical work (up to his Adam's Prize essay on Saturn's rings) was either pure geometry or very geometrical in its style. His mathematical vocabulary did broaden out thereafter, though the geometrical approach always remained his instinctive initial assault on any new topic. Throughout his career he continued to produce 'pure' geometrical papers—though they tended later to deal with topics thrown up by his work in physics, such as methods of tackling problems encountered in electromagnetism.

His first piece of physics was *On the Equilibrium of Elastic Solids* published in 1850.[1] It is a remarkably solid and complete piece of work for a teenager (Maxwell was then an undergraduate at Edinburgh). It contains an analysis of the 'stretchability of solids, based on the realization that their 'squashability' and 'twistability' may be quite different: Maxwell makes the comparison between jelly (basically water solidified with a little starch) which is very difficult to compress bodily, but has no resistance to twisting (shearing), and cork, which is much more compressible than water, yet on the other hand resists shearing forces much better also. Stokes and others had previously derived equations equivalent to Maxwell's, but he confidently set up his equations in his own personal notation, and then solved them to find the internal strains for a number of solids of various shapes subjected to a variety of mechanical tortures. Then he checked the accuracy of his mathematics by using polarized light as a strain gauge.

If white polarized light is passed through a transparent solid, the internal stresses reveal themselves as coloured interference bands seen in the solid. The effect is commonly observed by looking at the pre-stressed safety windscreens of cars while wearing Polaroid sunglasses. Here the stresses are deliberately built in so the windscreen 'frosts' rather than shatters if hit by a rock, but the technique developed by Maxwell is actually very useful to engineers looking for unwanted stresses in their structures. A plastic scale model of the structure in question is built, and studied with polarized light. Where there are lots of tight bands of colour, there is also stress, and that is where mechanical failure is likely to occur. By studying plastic scale models of Gothic cathedrals, 20th-century engineers have at last been able to solve the mystery how 14th-century monks managed to erect their wonderful (and amazingly stable) constructions.[2] Maxwell applied the technique to the simpler (and calculable!) case of the strains in a transparent ring. He poured hot water with some isinglass dissolved in it between two cylindrical walls, and when it had set, watched how the interference fringes moved as the walls were twisted. He also obtained prestressed isinglass by twisting the walls while the isinglass was setting.

The discovery of such stress-induced interference fringes had been made by Sir David Brewster, in Edinburgh, in the early 19th-century, but Maxwell was the first to develop it from a party trick to the accurate experimental tool it has now become. Maxwell was always very proud of this work, and displayed hand-coloured plates of the patterns he had obtained by the fireplace at Glenlair.

The paper is interesting for a number of reasons; firstly it is always pretty science to predict and then experimentally to verify an effect, and Maxwell here showed his early maturity and the balance of his theoretical and practical talents. He also shows his self-confidence in developing his own notation, deciding it was the best and sticking to it throughout the work. His panoramic vision of science is shown by the range of practical applications he made of his new technique: he applied it to calculate the bending of a silvered circular glass plate stretched like a skin over an evacuated iron 'drum'. Someone had suggested using such a device to make an astronomical telescope; Maxwell pointed out it would make a rather better aneroid barometer.

More than this, though, the paper sheds light on his subsequent work. He here developed from first principles the theory of nonisotropic solids (those whose properties in different directions are different, for example, isinglass with its different response to compression and shearing). His later molecular vortex aether was just such a solid, and the knowledge and mathematical confidence developed in the first paper must have been invaluable in the later one. Moreover, one of the special cases he especially considered in the first paper was just the effect of centrifugal forces on grindstones and flywheels rotating at speed. He also mentions, but did not attempt to deal with 'imperfectly elastic bodies'.

If a 'solid' is given a gentle and rapid shove, it will deform, but regain its original shape when the force is released: this is what is meant by a solid. Any solid exposed to a sufficiently large force, however, will become permanently squashed; steel girders can be bent and buckled into new shapes. Under these circumstances, the metal can be said to have flowed into a new pattern, and for the duration of the force, the solid acted as a fluid. The difference between solids and fluids is then largely one of degree: fluids flow under 'normal' circumstances, solids under abnormal ones.

As Maxwell pointed out, fluids (such as isinglass, i.e. almost pure water) are as incompressible as solids. They differ in being twistable (having very low shear resistance), so Maxwell's theory was based on characterising all solids by two parameters, a compressibility and a twistability, which varied gently with the shear being applied, up to a certain limit when the solid's resistance to twisting dropped to zero, its twistability became infinite, and it flowed like a fluid.

In the molecular vortex model of the aether, Maxwell began (in the first two parts of the paper) by considering the aether as a fluid: when an *electrical* force was applied to the idle wheels they moved, freely and easily in a wire to produce a current, but sluggishly in the aether:

> We know enough to be certain that the conducting powers of different substances differ only in degree, and that the difference between glass and metal is, that the resistance is a great but finite quantity in glass, and a small but finite quantity in metal.[3]

The idle wheels in a copper wire flow like water, in a dielectric (like glass) or in the aether, they flow like pitch or tar. The viscosity of the aether is high.

The insight that led to parts 3 and 4 of the molecular vortex paper, and to the theory of light, was to reverse the analysis of the 1850 paper. There was another way that a substance could respond to a force apart from flowing: elastic strain. By analogy with what he already knew for material substances, Maxwell realized that the aether too should be capable of an electrical strain as well as a current, and for short, sharp vibrating forces, below the elastic limit of the medium, it would be the elastic effect that would dominate, the aether was a visco-*elastic* medium and, like pitch, would appear solid over short periods of time.

The relationship between shear elasticity and viscosity remained a central concern for Maxwell. In his 1866 paper on gas theory, he was concerned with the calculation of the viscosity of a gas, based on the fundamental postulate of atomic collisions, but the paper begins with a clarification of how the coefficient of viscosity for a gas is related to its macroscopic shear-resistance, the quantity with which he had been concerned in 1850. The two are related through the relaxation time, T, the time the fluid takes to accommodate itself to a static applied torque. Maxwell noted however, that

> In mobile fluids T is a very small fraction of a second, and E [(the coefficient of shear elasticity)] is not easily determined experimentally.[4]

It took him 7 more years to find an experimental proof that there was such an elastic coefficient. Back in 1850 he had used polarized light to measure strain in jelly—a solid that was nearly a liquid. In 1873, he once again used polarized light to measure strains in Canada balsam, a viscous fluid that is nearly a solid. Viewing the liquid in polarized light, he saw coloured fringes trailing the edges of the spatula he used for stirring; the balsam was being stressed by the stirring and became doubly refractive (like the glass in a windscreen), but being a fluid it relaxed to homogeneity very quickly, and so the fringes could be seen only very close to the edge of the spatula as it trailed through.[5]

Similar to the way the 1850 paper on elasticity led on to his 1861 electromagnetic paper, so another early paper, produced this time when he was a fellow at Cambridge, contains the seeds of his 1864 electromagnetic paper. Optical instruments had till then always been considered as assemblages of particular lenses and mirrors, but Maxwell introduced a new way of looking at such instruments. He dug up, revived and developed an old theorem of Roger Cotes (see p.135) to consider instruments as being 'black boxes'. It is irrelevant what happens inside the instrument, what is important is what it does to the beams of light entering the apparatus. Maxwell gave a minimum qualification for the 'black box' to qualify as a 'perfect' optical instrument—and derived for the first time (because nobody had bothered to do so before) the formula for the longitudinal magnification of an optical instrument.[6] The idea of treating an optical instrument (telescope, microscope or whatever) as a black box— ignore its guts and just look at what outputs it gives for what inputs—is exactly the Lagrangian, hypothesis-free approach he adopted to electromagnetic theory in his later work.

In the 1870s Maxwell returned to optical instruments to apply formal mathematical Lagrangian methods to them. William Rowan Hamilton, the inventor of the quaternion, had also developed a fresh approach to mechanics, a development of the Lagrangian method, and, with Tait, had applied it to optics (the Hamilton characteristic function is a path integral formulation of optics, technically (see p.103)). The method was mathematically extremely difficult, and few people other than Maxwell bothered to look at it. Later on people found a simpler way of handling the mathematics which has developed into the 'eikonal' technique.

This interest in instruments formed just one part of Maxwell's work in optics. Altogether he published more papers in this area then in any other subject on which he worked, his main interest being in colour vision. That interest started with his visit to Nicol's laboratory (where he obtained the trusty Nicol prisms he used as polarizers in his work on strain measurement) and his work in Forbes' laboratory. Together with Forbes he studied the physiology of colour vision, and they established the 3-receptor theory.

The human eye can detect the brightness of a source of light, the colour of that source, and a third quality, the 'purity' of the colour (how muddy a brown is, the difference between pastel and primary shades, the fact that purple, a non-spectral colour is seen as a separate colour). The detectability of these three qualities implies the existence of three different types of colour receptor in the retina of the eye. Thomas Young right at the beginning of the 19th century had been the first to realise this, though his work has been largely ignored. Maxwell also showed that Newton's work on colour vision was consistent only with a 3-colour theory, though Newton had not realized this himself.

Forbes and Maxwell found an accurate and utterly convincing way of proving the fact. They took a spinning top, and placed circles of coloured paper (obtained from D R Hay—the man who had inspired Maxwell to look at ovals (p.37)) on its top, with slits along a radius, so that the different discs could be pulled through each other at will, to make a composite disc exposing chosen segments of the different colours. When the top was then spun, the colours fused to a blur, and the resultant colour could be parametrised by the amount of each coloured sector exposed. In this way they were able to analyse different colours. To begin they tried mixing blue and yellow to get a green, but found this did not work, so they tried red and green together and got a splendid yellow resultant.

They realised that the colour receptors in the eye, although responding to some extent to all colours, each have a preference; one for red, one for blue and the third for green. The usual theory of colour mixing, that blue and yellow make green applies to colour subtraction, the process that goes on in painting. Maxwell found a neat demonstration; he shone white light through a yellow solution, which absorbs all the colours but yellow which it transmits. This light was then passed through a blue solution, one that extracts all but the blue from white light. Since neither of the solutions (or paints in common use) produce an absolutely *pure* colour, the blue fails to absorb some of the green as well as the blue, and the yellow fails to absorb some green also. When the two solutions are put in sequence, it is the green they both *fail* to absorb that is finally transmitted. The blur of colours obtained when the top is spun is, on the other hand, the *sum* of the colours reflected by the individual sectors of paper; a process of colour addition rather than colour subtraction.

Using the top, Maxwell and Forbes found they could match all colours with red, green and blue discs of paper; therefore there were three types of colour-receptor in the eye, especially sensitive to these shades, and which built up physiological (as opposed to spectral) colour. Young, in his neglected work, had reached the same conclusion, Maxwell discovered. Because of illness, Forbes was forced to give up these experiments at this point, and Maxwell continued alone to collect quantitative data on colour vision. He arranged his coloured-paper discs to sit on top of larger discs of black and white paper. The coloured discs were set to produce a grey tone, which he could balance against the grey obtained by covering different amounts of the white disc with the black one. He showed that the balance obtained in sunlight needed rebalancing when illuminated by the glare of gas light: the yellow tone of the gas lamp affected the way the eye saw the different colours, thus proving that colour vision was dependent on the eye and light source, and not intrinsic to the object observed.

Maxwell suggested a neat verification of his theory: if three photographic plates were taken of the same scene, but one through a red filter, the next through a green one and the third through a blue filter, then if the plates were developed and projected onto a screen with the separate colours shone through the appropriate plates, a full-colour rendition of the original scene should be seen by the eye (even though it was actually seeing the view in only three individual tones). This is the principle of colour photography.

Maxwell made the suggestion in a paper in 1855,[7] but it took six years for him to try it out. On May 17th 1861, he gave a lecture to the Royal Institution, where he exhibited three slides of a tartan ribbon, which had been taken through separate filters by Thomas Sutton, the editor of the journal *Photographic Notes*. Maxwell projected them onto the screen and the stunned audience saw the separate tones combine to a full-colour, glowing image of the original.[8]

It took a century before anyone could understand how Maxwell had done it. Not the basic principles of course, for 3-colour photography soon ceased to be a laboratory curiosity and became a multi-million dollar industry, but how Maxwell's experiments had worked. He was not using modern colour film, but the old-fashioned wet-collodion process. Today's print films are made relatively insensitive to red light: this enables red safety-lights to be used in darkrooms, but the wet-collodion process is totally impervious to red. Nothing should have come out on the plate taken through the red filter; yet enough did emerge for the demonstration to be convincing. In 1960 a team of scientists from Kodak explained the trick. The red filter Maxwell and Sutton used absorbed all visible light but the red, but it also transmitted some ultraviolet, invisible light. The red dye in Maxwell's tartan ribbon also reflected some ultraviolet light, at exactly the wavelength the filter let through. Thus by a double stroke of fortune, Maxwell's red was actually ultraviolet. Ultraviolet light is very energetic light, of shorter wavelength than blue; it causes tanning and sunburn and it does expose wet-collodion. The clincher for the Kodak team came when they noticed that the original red plate was slightly out of focus—exactly as it should be if it was focussed for red light, but was actually exposing the picture by ultra-violet.

Maxwell continued his 1855 colour vision paper by looking at colour blindness. The great chemist Dalton had been colour blind (colour-blindness is sometimes called Daltonism) and on one memorable occasion had embarrassed all his acquaintances by turning up at a funeral wearing a bright red suit. John Herschel, knowing of Young's theory of colour vision had once written to Dalton suggesting that maybe the red receptors in his eyes were not working properly. Maxwell was able to prove this hypothesis to be true, not for Dalton, but for colour-blind volunteers, whom he got to try and balance colours: they always thought they had obtained a balance using only the blue and green coloured discs. In Aberdeen he 'volunteered' the students attending his lectures for the colour-top tests. The next day, he noted to Tait, the one student who had discovered he was colour-blind absented himself. Through genuine illness it transpired.

Maxwell devised a method for the colour-blind to 'cure' themselves, derived from his daylight/gaslight comparison. He made up for them spectacles with one red and one green lens: these filtered the incoming light quite differently, so that two shades that might appear colour-balanced through one lens would not match through the other. This provided the wearer with enough extra information to make up for his missing receptors. Unfortunately there is a snag. Though the red and the green lens were indistinguishable to the colour-blind person—they both appeared uniformly dark—the social stigma attached to walking through life as if one had just emerged from a 3-d movie has prevented the idea from ever catching on.

The colour top which Maxwell used was a delightful toy, and he developed it into the 'dynamical top' a much more formidable instrument: a 2¼ lb brass bell balanced on a steel point, and bristling like a dinosaur with brass screws. These could be adjusted to alter the dynamic balance of the bell (just like car wheels) and so when the top was spun it could perform all sorts of gyrations. It was intended to provide a visual, mechanical demonstration of the intricacies that could arise in the rotation of asymmetric bodies.

To continue with his work on colour vision, however, Maxwell found the colour top difficult. It was almost impossible to standardise the colours of the paper discs used by different experimenters (impossible, in fact, to get a printer to print successive batches of discs the same colour). Maxwell therefore developed the 'colour box', basically a large spectrometer, where sunlight was admitted at one end through three slits adjustable in both width and position so that different intensities at different wavelengths could be sent to the final focus (thus replacing the different sized sectors of differently coloured paper). The balance was achieved against ordinary white light, admitted unaltered through to the focus by a different route.

With this device, through its various modifications, Maxwell carried out many hundreds of tests of people's ability to match colours. What he convincingly showed was that though different people see slightly different colours, for each individual, the eye can match colour with remarkable accuracy and replicability. With care he obtained accuracies of 0.2%.

His correspondent Monro attempted to measure to 0.1%. Maxwell advised him that it was very difficult, and added 'if you can get observations to be consistent to the 3rd place of decimals, glory therein, and let me know what the human eye can do'.[9] Even 0.2% is far more accurate than the hand's ability to measure distance (see p.47).

One of Maxwell's optical discoveries relates to the 'yellow spot' of the retina. This is a central region where the retina has a yellowish colour (i.e. it absorbs blue-green light). Maxwell found that it acted on incoming light to chop the eye's sensitivity to a particular narrow blue-green band of colours very sharply. Colour matching experiments done in this part of the spectrum using this part of the retina then revealed how yellow were the yellow spots of different people. Maxwell had a very yellow yellow-spot, and his measurements for this blue-green region were rather inaccurate. Katherine Dewar his wife-to-be, who, as the observer K, was assisting him in these experiments, turned out to be very unusual in that she seemed to have no yellow in her yellow-spot at all; the balances she obtained for this region of the retina matched well with those she obtained for off-axis regions.[10]

Maxwell found a test for the existence of the yellow spot in a person's retina, the Maxwell spot test. One form of it was to look at a screen illuminated by white light through a solution of chromium chloride. This allows through only a band of red light, and the blue-green band strongly absorbed by the yellow spot. Together the two colours give a pretty neutral mix, which is seen to illuminate most of the screen. But at the centre of the field of vision, the yellow spot allows only the red light through, and so a red spot 'floating like a rosy cloud' is seen there.

Maxwell also applied his geometrical (and analytical) skills to problems in engineering and thermodynamics. In engineering he developed one of Rankine's ideas, to replace the structural diagrams of frameworks by their reciprocal force diagrams, and went on to apply this to continuous media. He was awarded the Keith Prize by the Royal Society of Edinburgh for this work.[11]

In thermodynamics, he wrote an 'elementary' introduction to the subject *The Theory of Heat* which nonetheless contained a very important original piece of work. By a geometrical argument involving a complicated mess of construction lines on a pressure/volume graph, he derived a set of equivalent relations between the pressure, volume, temperature and entropy, which he expressed as four differential equations: Maxwell's equations (or relations, as they are sometimes called to avoid confusion).[12] In later editions of the book, Maxwell included a section on the work of the American physicist Josiah Willard Gibbs, who in 1875 proved some powerful theorems on phase equilibrium and how substances pass from the solid to liquid, and liquid to gas phases.[13] Maxwell had, himself, almost simultaneously derived identical results, as a letter to Stokes written in the summer of 1875 reveals,[14] but he gave all the credit to Gibbs, the foundation of his own proof rested solidly on Gibbs' earlier work. Maxwell then was able to visualise Gibbs' results geometrically in a 3-dimensional space with volume, entropy and energy as his axes. The behaviour of any substance can be represented by a particular surface in this space, and Maxwell constructed the one for water, which is still preserved in the Cavendish Laboratory. Gibbs' results can be derived, Maxwell wrote, by sliding a pane of greasy glass over the surface contours of the model.

Maxwell felt a deep admiration for Gibbs' work. In a letter to Tait he wrote 'he has more sense than any German',[15] somewhat unfairly, and there is a story that the president of Yale University came once to Cambridge to ask Maxwell to suggest an eminent European physicist whom Yale could appoint as professor to build up their scientific reputation. 'What do you want a European for, when you already have Gibbs' he was told. Yale's president, the story goes, had never heard of Gibbs, and was unaware that he was already employing him.

The final aspect of Maxwell's work are his 'ephemeral' papers—one-off pieces which grew out of everyday life and observations. He did not regard physics as office work, to be forgotten on leaving the laboratory. He was concerned with understanding Nature, which he regarded as all of a piece, and always testifying to the ingenuity of its creator. He maintained a constant delight in the marvels of the natural world which enveloped him. He had the passion of the amateur.

His paper on the fish-eye lens has already been mentioned (p.61). Another delightful oeuvre from his early Cambridge days dealt with dropping pieces of paper. If a rectangular piece of paper, $2'' \times 1''$ is best, is dropped in still air, it will make a few initial flutters and flurries before settling into a steady descent to the floor, inclined at a constant angle to the vertical, rotating rapidly all the while. Maxwell was able to find a simple explanation (he uses no mathematics at all in the paper) of the steady part of the descent. One cannot help feel the paper was inspired by Maxwell making some false starts or silly arithmetical mistakes in tackling Hopkins' problems; most people in such circumstances scrumple up the soiled paper, or, more constructively, convert it into a paper dart. Perhaps Maxwell let one or two sheets (bigger than the ideal $2'' \times 1''$, that would smack of hubris even for Maxwell) drop to the floor, observed the fluttering, and converted his mistakes into an original piece of research.[16]

Another neat little paper may also have resulted from a mistake: to obtain nice spectra he had to use a prism in combination with an achromatic lens. But what happens if one forgets one's lens? Maxwell wrote a paper on a theoretical combination of prism and ordinary lens that would avoid the problem. It had the added advantage of being applicable to infra-red spectra, where achromatic lenses were not available.[17]

When Maxwell was doing the resistance standardisation experiments at King's (p.79) the big coil had to be kept rotating at a constant speed, for long periods at a stretch. To help the experimenters maintain the constancy of the rate of revolution they used the Thomson-Jenkin Governor. Governors are, in general, devices designed to keep machines running automatically at a constant rate. If they start accelerating, for example, a value might cut down the fuel supply, if they run too slow, the same valve might open up the fuel supply. The first such device had been invented by James Watt, to keep his steam engine working steadily and efficiently. Since Watt's time, many other governors had been produced. Maxwell thought about the different designs, (perhaps while idly keeping his coil turning, as the coil was in this case hand-cranked) and wrote a paper On Governors, about the theoretical principles of self-regulating machines. Norbert Wiener, who developed the theory of such machines into the science of 'cybernetics', regarded Maxwell's paper as the foundation of the subject, and invented the name cybernetics from the same Greek word as was derived 'governor'.[18]

In his construction work at Glenlair, and later on when building the Cavendish Laboratory at Cambridge, Maxwell became interested in the protection of buildings from lightning. He became recognized as something of an expert on this subject, and when Tait moved into a big new house in Edinburgh, he promptly wrote to Maxwell to ask for advice 'Answer speedily, as I have a house to protect'.[19]

Maxwell replied the following day, saying that the Provost of King's College had also asked him for advice (see p.76), and he also produced an Encyclopaedia Britannica article on the subject. Lightning conductors make a fascinating study, for they display quite beautifully the difficulties inherent in applying simple and straightforward physical principles to the complexity of Nature.

It had long been thought that lightning conductors worked by drawing the 'sting' from approaching thunderclouds. As was well-known, charge leaks away from or onto sharply-pointed objects much more easily than from smooth ones. This is now understood as happening because the very intense electric fields in the neighbourhood of a sharp point rapidly accelerate to very high speeds any charged particles that happen to be in the air there, so rapidly that when these particles collide with uncharged molecules (as the kinetic theory of gases says they must, sooner rather than later) they have enough energy to break the neutral molecules apart into positively and negatively charged fragments. These are also accelerated by the intense fields, and repeat the process. There is a chain reaction breakdown of the elctrical resistance of the air. If an electrically charged thundercloud approaches a pointed conductor, the voltage difference between the two produces these intense fields and the atmosphere between the cloud and the point no longer acts as an insulator, but instead conducts a steady discharge of the thundercloud. Thus, went the theory, the purpose of the lightning conductor is to cause a slow steady earthing of the thunderstorm, rather than the sudden disruptive discharge of a lighting stroke.

Maxwell disagreed

> It appears to me that these arrangements are calculated rather for the benefit of the surrounding country and for the relief of clouds labouring under an accumulation of electricity, than for the protection of the building on which the conductor is erected.
>
> What we really wish is to prevent the possibility of an electric discharge taking place within a certain region, say in the inside of a gunpowder manufactory. If this is clearly laid down as our object, the method of securing it is equally clear.[20]

Maxwell proposed that a lightning conductor should effectively enclose a building in a Faraday cage. If a lightning conductor *grid* was put around the building, made of a good conductor such as copper wire, then it would effectively keep the outside of the building all at the same voltage. It is then a consequence of Coulomb's law of electrostatic force that there could exist no voltages inside the building, and so there could be no sparks (see p.191).

The problem with keeping the building as an equipotential cage was that water and gas pipes leading into the building were connected to earth, and so to avoid sparks between the lightning conductor and the piping, the two should be connected. In his reply to Tait's request (which he addressed O(T + T')–T) he wrote

> In a town there is 0 like waterpipes. Gas pipes are jointed with white lead and are not good conductors and are also fusible.[21]

The difference between Maxwell's and the then current view was that instead of pointed lightning conductors to encourage a constant trickle of current down the outside of the building, Maxwell thought it unimportant whether the building was hit by a lot of little strokes of lightning (effectively) or one big one. The critical thing was that the whole of the building be effectively covered with a net of conductors so that there be no sparks inside the building. To him it was irrelevant whether the conductors' tops were pointed or not.

If Maxwell's views succeeded in percolating through to the Fellows of King's, they singularly failed to do so to Tait, for in a lecture given later, on 29/1/1880, Tait reiterates the standard view of lightning conduction.[22]

Actually, both Tait and Maxwell were wrong. Tait because the spikes on the tops of conductors are negligible compared to the spikes Nature provides, on twigs, blades of grass, etc., whose effect must outweigh that of the conductor. Maxwell was wrong because he failed to take into account his own equations of electromagnetism. His views on lightning conduction would work fine if only the static potential difference was involved. But in a lightning strike time also comes into play, and then there is another term that can cause those dangerous sparks: a varying magnetic vector potential. A sudden surge of current down a lightning conductor will induce voltages inside the building through the varying magnetic field it creates. The application of theory to practice is ever fraught with pitfalls.

Maxwell also got involved in the ball lightning dispute, a vitriolic old chestnut even then. In a letter to Thomson he wrote with unconcealed delight:

> I enclose an interesting account of Mr. and Mrs. Brown of Charles St., Windsor, particularly the latter, whose patient observations of the fireball deserve all praise . . . Mrs. Brown on a rough estimate, is of opinion that the fireball must have contained at least a farad of the electrical fluid. Its potential at the intstant of striking the chimney stack must have been several mega volts. Mrs. Brown, who with more than the lightning's speed pursued the fiery globe, had probably a resistance of not more than 1,000 Ohms at her command. Her courage, therefore, in making herself mistress of all the details of the phenomenon deserves a public recognition by the British Association.[23]

Winter conditions at Glenlair afforded him more scope for natural philosophy. The marvellous eerie sound of curling stones on ice attracted his attention. Because the noise comes from the vibration of a large area of ice, it is not particularly loud if you stand on top of it, but equally the sound carries for miles, as he wrote to his scientifically-minded correspondent Droop.

> We have clear hard frost without snow, and all the people are having curling matches on the ice, so that all day you hear the curling stones on the lochs in every direction for miles, for the large expanse of ice vibrating in a regular manner makes a noise which, though not particularly loud on the spot, is very little diminished by distance.[24]

On another occasion he saw a rainbow on the surface of a frozen ditch at Cambridge, 'I at once made a rough measurement of the angle on the board of a book I had with me'. Maxwell attributed the bow to small almost spherical droplets of water sitting on the surface of the ice. 'The ice was very thin, and I was not able to get near enough to the place where the bow appeared to see if the supposed water drops really existed' he confessed to the Royal Society of Edinburgh.[25] Scientific dedication goes only so far.

In summarising his 'other work' we repeat the words of Professor Charles Coulson, who noted that although the total number of papers published by Maxwell is not large,

> There is hardly one of them which does not open up some new field. There is scarcely a single topic that he touched upon which he did not change almost beyond recognition.[26]

CHAPTER 13

Maxwell was Cambridge's third choice to be the Cavendish Professor of Experimental Physics. Only after Thomson in Glasgow, and Helmholtz in Berlin had turned the job down was if offered to Maxwell. Cambridge was luckier than it deserved. It got the greatest of the three very great scientists.

As the 19th century progressed, it became inevitable that proper science teaching would have to be brought in. Science was moving away from the days when the amateur playing with his candles and string in his country rectory could make great discoveries (Maxwell realised he belonged to an old and dying school). Precision apparatus, able to subject matter to extreme conditions was becoming the norm: vacuum pumps and very large electo-magnets were standard laboratory equipment to which only the exceptional amateur could aspire. To use such devices required training, and there was a move towards training undergraduates in laboratory technique.

Thomson had begun it in Glasgow, when, in the 1840s he took over an old wine cellar, removed the racks (hopefully he did not waste their contents) and installed some apparatus. But here it was keen undergraduates assisting the Professor in his experiments—Thomson took out the patents in his own name.

In Edinburgh, Maxwell had similarly had the run of Forbes' own laboratory, and when he himself became professor at Aberdeen and King's he instituted laboratory teaching for his students. At King's indeed, there may already have been something of a tradition of such teaching. When the Civil Engineering Department was set up in 1839, it was intended that

> The observation, judgement and invention of the students will be exercised by experiments made by themselves, and by visits to various manufactories and other works, to which access has been liberally granted . . . and where they will be accompanied by the lecturer, who will give explanations on the spot.[1]

Wheatstone presumably would have taught these engineers electricity, and there is some evidence that he may even have got them to make electric machines for him. Engineers were in a rather different position from natural philosophers, however, theirs was an essentially vocational training—very few engineers went to university at the time, almost all (including Edward Harland, see p.31) were apprenticed directly from school.

However, once Oxford had decided that it needed proper laboratories for its physicists, in 1868, it was inevitable that Cambridge would bend with the prevailing wind.

The need felt for such teaching is testified to by Lord Rayleigh. His wife once wrote in her diary

> Rayleigh has been talking this evening about the difficulty of beginning his experimental career. It was not only that his tastes were entirely out of the range of his family—there was no one at Cambridge to give him a helping hand, or suggest his going to Germany, or ask him what he intended to do. There was no one experimenting there except Stokes and Liveing. Stokes' lectures interested him immensely, and he was kind in answering questions, but it never seemed to occur to either of them to try and help a young man forward. John got the idea from something that he read that he would like to work at volumetric analysis— but someone (a sort of laboratory assistant I think) could not imagine what he should want to do that for . . . and thus, though he stayed up to read, 3 precious years of experimental work were lost for want of knowing how to set to work. Glasgow? He had never even heard that you *could* work there.[2]

Of course, not all the undergraduates would progress to scientific careers, but nonetheless, to have scientifically educated people in all walks of public life was then becoming vital. A university commission was set up, and decided in 1869 that Cambridge must have its own laboratories. The money needed—an estimated £6,000—was frightening, for those days, and the verdict was by no means unanimous. No less a figure than Isaac Todhunter, the Wrangler-maker of the day, admitted the value of *original* experiments, but doubted whether their repetition brought much enlightenment 'he who first plucks an experimental flower thus appropriates and destroys its fragrance and its beauty.' This may not be altogether untrue, but he continued with these immortal words:

> . . . It may be said that a boy takes more interest in the matter by seeing for himself, or by performing for himself, that is, by working the handle of the air-pump; this we admit, while we continue to doubt the educational value of the transaction. The boy would also probably take much more interest in football than in Latin grammar; but the measure of his interest is not identical with that of the importance of the subjects. It may be said that the fact makes a stronger impression on the boy through the medium of his sight, that he believes it the more confidently. I say that this ought not to be the case. If he does not believe the statements of his tutor—probably a clergyman of mature knowledge, recognized ability and blameless character—his suspicion is irrational, and manifests a want of the power of appreciating evidence, a want fatal to his success in that branch of science which he is supposed to be cultivating.[3]

With that sort of antagonism to the idea of laboratory teaching, the University may well have put their Commission's expensive recommendation into the permanently-pending tray, but they were then made an offer they could not refuse: the Chancellor of the University, the Duke of Devonshire, offered to pay for the laboratory out of his own pocket.

William Cavendish, the Duke of Devonshire, was a remarkable man; descended from Henry Cavendish, the great 18th century physicist, he had, like Maxwell, been 2nd Wrangler and 1st Smith's Prizeman in his year, but gone in for a political career (a proof of the Tripos' educational efficacy) before becoming Chancellor of the university. In 1872 he was made chairman of the

Devonshire Commission, a Royal Commission to look into the relations of the State with science, to see if more state support was needed. The commission uncovered some remarkable facts: at Oxford there were 9 Natural Science fellowships out of 165, at Cambridge there 3 out of 105; Britain was producing about one sixth of the number of chemistry research papers as Germany; the Government was spending more on its Natural History *collections* than on all other branches of science together. The Commission recommended increased government support, that there should be broad educational courses at the universities without early overspecialization, and that there should be a radical reform of school education. The proposals were, by and large, ignored.[4]

The Duke had decided, however, to reform his own university off his own bat, and his generous offer meant that they could now proceed with their plans. Towards the end of 1870, it became known that a chair would be established the following February, and overtures were first made to Thomson. He replied that the Cambridge climate and the congeniality of life there both had great appeal, but his wife had just died and he now felt that all he wanted to do was concentrate in peace on science for the rest of his days (which were to number 37 more years and include a second wife and a peerage), he had his university life comfortably organized in Glasgow, and, significantly, mentioned 'the convenience of Glasgow for getting mechanical work done'. He rejected the offer.[5]

Thomson was asked if *he* would sound out Helmholtz, but Helmholtz had just been appointed the head of the new Institute of Physics in Berlin and was happy there. With these two out of the running, the body of opinion swung to Maxwell. The Hon. J.W. Strutt (later Lord Rayleigh, and Maxwell's successor in the Cavendish chair) wrote to him.

> When I came here last Friday I found every one talking about the new professor-ship, and hoping that you would come. Thomson it seems, has definitely declined and there is a danger that some resident may get promises unless a proper candidate is soon in the field. There is no one here in the least fit for the post.[6]

Maxwell received another letter the same day from the Reverend Blore containing the same message—but politely saying that he mentioned Thomson's refusal only in case Maxwell might not wish to stand against him.

Maxwell's first impulse was to decline

> My dear Blore—Though I feel much interest in the proposed Chair of Experimental Physics, I had no intention of applying for it when I got your letter, and I have none now, unless I come to see that I can do some good by it.[7]

but the way he continued the letter must have given Blore good ground to hope that he was persuadable (though Campbell fails to quote this bit in his biography). Maxwell asked Blore a series of very detailed questions; what are the duties? is the professor attached to a college? who appoints him? is it for life? how many terms a year does he have to keep? are the students to do experiments? who are the *other* candidates?

Maxwell was persuaded to stand with the understanding that if elected he might choose to resign at the end of the year. There were no governmental crises this time, and also no opposition for the chair. Maxwell was elected on March 8th.

He was now the Professor of Experimental Physics and had no laboratory, but this was an advantage, for he was able to specify to the architect what facilities he thought a laboratory should have, based on his own experience, and a series of visits he now paid to laboratories about the country. The building he and the architect produced was to work remarkably well. It was comfortable, unostentatious and practical enough to house Cambridge physics for the best part of a hundred years.

Some of the design features are especially noteworthy. The magnetic disturbances caused by Thames steamers were not a worry in rural Cambridge, but traffic vibrations could be. A special magnetic room was built with three monolithic piers set on their own foundations through holes in the floor: sensitive apparatus could sit on their tops. Also on the ground floor was a workshop, for making apparatus and a battery room to supply the building with electricity. At the top of the building's tower, 50 feet up, was put a large water tank, to get up enough pressure to drive a big vacuum pump. Vacuum lines were run to the various rooms in the laboratory so that students could, in the italics of *Nature*'s observer at the opening ceremony, *turn on a vacuum*. Another neat device was that all South and East facing windows had stone sills 18" wide, so that heliostats could be placed there—mirrors turning to follow the sun to channel sunlight into the laboratory. Sunlight (when available) was still the only strong source of steady illumination. All these features show a very good grasp of the practical necessities of laboratory design.

The lowest tender for the building was £8,450, £2,000 over the original estimate, but the Duke agreed to meet this and further furnish the laboratory with all necessary apparatus. Maxwell drew up a provisional list of requirement, and also brought his own private stock of equipment, and the .B.A donated the apparatus Maxwell had had built for the electrical standards experiments. Adequate furnishing of the laboratory could not be completed on the spot, and Maxwell realised that a steady trickle of apparatus would have to be bought over the years. Rather than pester the Duke with persistent requests, but not wanting to deprive him of the feeling that the Cavendish, lock, stock and barrel was his gift, Maxwell reported to the University that 'The Chancellor has completed his gift to the University by furnishing the Cavendish Laboratory with apparatus suited to the present state of Science'. This statement conceals the fact that the 'present' state of science is continually changing and a laboratory needs apparatus to explore possible 'future' states of science too. Maxwell reserved for himself the 'liberty' to buy extra equipment when needed, and during his tenure of office he spent several hundred pounds of his own money. Even so, when Rayleigh succeeded him in the Chair he found the stock of apparatus woefully inadequate: knowing that the professor was paying for equipment from his own pocket made many of the reasearchers reluctant to approach him for any but the most essential apparatus. Some of the techniques they were forced to use sound almost comical. To do an experiment that required a dry atmosphere, a well-dried woollen blanket would be hung in the laboratory. It would often gain two pounds weight in twenty-four hours. This was obviously no way to encourage fundamental research, and is perhaps indicative of Maxwell's already old-fashioned patrician, amateur approach to research. Rayleigh, far more of the practical businessman than Maxwell (he delayed announcing the finding of the new element argon so the the discovery could be entered for a prize of ten thousand dollars from the Hodgkins Fund in America for 'a treatise embodying some new and important discovery in regard to the nature and properties of atmospheric air'), immediately set up an apparatus fund for the laboratory. He gave five hundred pounds, and the Duke of Devonshire gave another five hundred.[8]

The actual building of the laboratory was begun in 1872 and finished late in 1874. The delay exasperated Maxwell,

> But at present I am all day at the Laboratory, which is emerging from chaos, but is not yet cleared of gas-men, who are the laziest and most permanent of all the gods who have been hatched under heaven.[9]

forestalling once again Flanders and Swan. His teaching duties had actually to start in October 1872, and this caused some early inconvenience:

> I have no place to erect my chair, but move about like the cuckoo, depositing my notions in the chemical lecture-room 1st term; in the Botanical in Lent, and in Comparative Anatomy in Easter.[10]

Maxwell's inaugural lecture was another magisterial resumé of the state of science, and his intentions for the new laboratory. It also had strong overtones of appeasement for those who had initially disapproved of the idea of the laboratory, but these were all wasted, because there had been only a casual announcement of when it was due, and all the university bigwigs missed it. Then the first of his academic lectures *was* announced formally, and all the university officers mistakenly attended that. The result was that Maxwell with a twinkle in his eye gave Cayley, Stokes, J.C. Adams and the rest a sober and very thorough explanation of the difference between the Centigrade and Fahrenheit scales of temperature. It was rumoured that Maxwell had not been entirely innocent in the affair.

Fortunately the true Inaugural Lecture was later printed and so did not waste its fragrance. It is a quite remarkable document. In it he took up Todhunter's criticisms of experiment; Maxwell did not pretend that all experiments were earthshatteringly important: he drew a distinction between experiments of research and those of illustration (almost exactly as he had done at King's):

> The aim of an experiment of illustration is to throw light upon some scientific idea so that the student may be enabled to grasp it. The circumstances of the experiment are so arranged that the phenomenon which we wish to observe or to exhibit is brought into prominence, instead of being obscured and entangled among other phenomena, as it is when it occurs in the ordinary course of nature. To exhibit illustrative experiments, to encourage others to make them, and to cultivate in every way the ideas on which they throw light, forms an important part of our duty.

The simpler the materials of an illustrative experiment, and the more familiar they are to the student, the more thoroughly is he likely to acquire the idea which it is meant to illustrate. The educational value of such experiments is often inversely proportional to the complexity of the apparatus. The student who uses home-made apparatus, which is always going wrong, often learns more than one who has the use of carefully adjusted instruments, to which he is apt to trust, and which he dares not take to pieces.

It is very necessary that those who are trying to learn from books the facts of physical science should be enabled by the help of a few illustrative experiments to recognise these facts when they meet with them out of doors.

In an experiment of research, on the other hand, this is not the principal aim. It is true that an experiment, in which the principal aim is to see what happens under certain conditions, may be regarded as an experiment of research by those who are not yet familiar with the result, but in experimental researches, strictly so called, the ultimate object is to measure something which we have already seen—to obtain a numerical estimate of some magnitude.[11]

Experiments of illustration are for undergraduates, the research is to be left to the professor and his graduate students.

Maxwell must have suspected that what Todhunter was really afraid of was that potential first class material would be seduced by the siren call of acoustics—and heat, electricity and magnetism

. . . If we succeed too well, and corrupt the minds of youth till they observe vibrations and deflexions, and become senior Ops intsead of wranglers, we may bring the whole university and all the parents about our ears.[12]

the other worry was that students might take on too much work and have nervous breakdowns. Maxwell volleyed these points away beautifully. He admitted the problem, but showed that because a lot of Tripos was applied mathematics, then knowing what it applied to would be a great help in the exam:

No doubt there is some reason for this feeling. Many of us have already overcome the intial difficulties of mathematical training. When we now go on with our study, we feel that it requires exertion and involves fatigue, but we are confident that if we only work hard our progress will be certain.

Some of us, on the other hand, may have had some experience of the routine of experimental work. As soon as we can read scales, observe times, focus telescopes, and so on, this kind of work ceases to require any great mental effort. We may perhaps tire our eyes and weary our backs, but we do not greatly fatigue our minds . . .

I quite admit that our mental energy is limited in quantity, and I know that many zealous students try to do more than is good for them. But the question about the introduction of experimental study is not entirely one of quantity. It is to a great extent a question of distribution of energy. Some distributions of energy, we know, are more useful than others, because they are more available for those purposes which we desire to accomplish . . .

There may be some mathematicians who pursue their studies entirely for their own sake. Most men, however, think that the chief use of mathematics is found in the interpretation of nature. Now a man who studies a piece of mathematics in order to understand some natural phenomenon which he has seen, or to calculate the best arrangement of some experiment which he means to make, is likely to meet with far less distraction of mind than if his sole aim had been to sharpen his mind for the successful practice of the Law, or to obtain a high place in the Mathematical Tripos.

I have known men, who when they were at school, never could see the good of mathematics, but who, when in after life they made this discovery, not only became eminent as scientific engineers, but made considerable progress in the study of abstract mathematics. If our experimental course should help any of you to see the good of mathematics, it will relieve us of much anxiety, for it will not only ensure the success of your future studies, but it will make it much less likely that they will prove injurious to your health. [13]

Maxwell also made an appeal to the arts faculties, that science was revolutionising life, and was far too important to ignore. Moreover it was an important element in cultural life

. . . we are daily receiving fresh proofs that the popularisation of scientific doctrines is producing as great an alteration in the mental state of society as the material applications of science are effecting in its outward life. Such indeed is the respect paid to science, that the most absurd opinions may become current, provided they are expressed in language, the sound of which recalls some well-known scientific phrase. If society is thus prepared to receive all kinds of scientific doctrines, it is our part to provide for the diffusion and cultivation, not only of true scientific principles, but of a spirit of sound criticism, founded on an examination of the evidences on which statements apparently scientific depend . . .

We are not here to defend literary and historical studies. We admit that the proper study of mankind is man. But is the study of science to be withdrawn from the study of man, or cut off from every noble feeling, so long as he lives in intellectual fellowship with men who have devoted their lives to the discovery of truth, and the results of whose enquiries have impressed themselves on the ordinary speech and way of thinking of men who never heard their names? Or is the student of history and of man to omit from his consideration the history of the origin and diffusion of those ideas which have produced so great a difference between one age of the world and another? [14]

Exactly the tenor of the Devonshire Commission proposals of later on. The great scientists of the day were all cultured men, who valued the breadth of education they had received and from which they had profited. Maxwell's determined efforts to keep up his Latin and Greek all his life have already been mentioned.

After the anticlimax of the opening lecture, Maxwell and the new laboratory settled down to work. A nucleus of reasearchers formed around Maxwell, some Cambridge graduates, others, like Arthur Schuster, attracted to work with the great man (he gave up a job at Manchester university to come to Cambridge). Hicks went on to be vice-Chancellor of Sheffield University, Chrystal Professor of Mathematics at Edinburgh, Gordon a scientific amateur and secretary to the B.A., Fleming left a post at Cheltenham College to work with Maxwell and went on to become professor of Electrical Engineering at Cambridge (and invent the triode), Maxwell suggested Shaw look at a problem for the Meteorological Council and from that he went on to become the Director of the Met. Office, Poynting (discoverer of the eponymous vector) became Professor of Physics at Birmingham, MacAlister Principal of Glasgow University . . .

Maxwell was not the dynamic leader of a research team, as in today's big science; more he regarded himself as responsible for a group of independent gentlemen sharing a roof. Schuster later described it thus

> Maxwell might easily have found students eager to work out in detail some problem arising out of his theoretical investigations. This would have been the recognised method of a teacher anxious to found a 'school'; but it was not Maxwell's method. He considered it best both for the advance of science, and for the training of the student's mind, that everyone should follow his own path. His sympathy with all scientific inquiries, whether they touched points of fundamental importance or minor details, seemed inexhaustible; he was always encouraging, even when he thought a student was on a wrong track. 'I never try to dissuade a man from trying an experiment,' he once told me; 'if he does not find what he wants, he may find out something else.'[15]

J.J. Thomson, the third Cavendish professor, wrote that this laissez-faire method was exactly his approach too. It did not of course mean that Maxwell just left his students completely alone. He suggested interesting problems for them to begin on (for it is one of the most difficult questions for a student starting out to know which areas are likely to prove interesting and remunerative), designed the intitial experimental approach, and always kept a paternal interest in the experiment's progress

> Maxwell often showed a certain absent-mindedness: a question put to him might remain unnoticed, or be answered by a remark which had no obvious connexion with it. But it happened more than once that on the following day he would at once refer to the question in a manner which showed that he had spent some time and thought on it. I never could quite make up my mind whether on these occasions the question had remained unconsciously dormant in his mind until something brought it back to him, or whether he had consciously put it aside for future consideration, but it was quite usual for him to begin a conversation with the remark: 'You asked me a question the other day, and I have been thinking about it.' Such an opening generally led to an interesting and original treatment of the subject.[16]

Another of his students, Glazebrook, who later became the Director of the National Physical Laboratory, also remarked on Maxwell's reliance on his subconscious

> When difficulties occurred Maxwell was always ready to listen. Often the answer did not come at once, but it always did come after a little time. I remember one day, when I was in a serious dilemma, I told him my long tale, and he said:—
>
> "Well, Chrystal had been talking to me, and Garnett and Schuster have been asking questions and all this has formed a good thick crust round my brain. What you have said will take some time to soak through but we will see about it." In a few days he came back with—"I have been thinking over what you said the other day, and if you do so-and-so it will be all right".[17]

The work Maxwell set his students was mostly a series of high-precision experiments. Perhaps the best example is the battery of careful tests to which the Cavendish team subjected Ohm's law, $V = i\,R$. The question was, just how linear is the relation between voltage and current, and the problem was that as the current increases so does its heating effect. Resistance certainly varies with temperature, so this effect has to be eliminated.

Schuster had done some work on this in Göttingen, and found that Ohm's law was not accurate. The BA thought this should be repeated, and George Chrystal took on the task. The method was to set up a Wheatstone bridge with arms made of two resistances of the same material, but with very different cross sections. The current in the bridge was rapidly alternated between a large and a small value, so rapidly that the temperature of the bridge remained effectively constant throughout. The apparatus was set to a null reading across the bridge. Then the direction of the weak current was reversed. Only if Ohm's law applied should there still be equilibrium, providing all possible complicating side-effects had been eliminated. The result: 'Ohm's law has come out triumphant, though in some experiments the wire was kept bright red-hot by the current' Maxwell wrote to Campbell.[18] More prosaically, Chrystal reported to the B.A. that the resistance of a conductor measured at 1 ohm for an infinitesimal current is altered by less than 1 part in 10^{12} (1,000,000,000,000) when carrying a current of 1 amp.

Maxwell wrote

> It is seldom, if ever, that so searching a test has been applied to a law which was originally established by experiment, and which must still be considered a purely empirical law, as it has not hitherto been deduced from the fundamental principles of dynamics. But the mode in which it has borne this test not only warrants our entire reliance on its accuracy within the limit of ordinary experimental work, but encourages us to believe that the simplicity of an experimental law may be an argument for its exactness, even when we are not able to show that the law is a consequence of elementary dynamical principles.[19]

It is interesting to compare this with remarks Maxwell made in a review of a book by Whewell which he wrote for *Nature*. Maxwell looked particularly at what is meant by 'mass' and 'conservation of mass'. Mass is never measured directly, weight is, and the weights of bodies must certainly change in a chemical reaction, because the weight of a body depends on where it is in the earth's gravitational field, and that must alter in a reaction. Nevertheless

> We are led by experiments which are not only liable to error, but which are to a certain extent erroneous in principle, to a statement which is universally acknowledged to be strictly true. Our conviction of its truth must therefore rest on some deeper foundation than the experiments which suggested it to our minds.[20]

Maxwell saw in the linearity of Ohm's law, as he had in gas viscosity, one more piece of evidence that the universe was created by a rational God.

In his Inaugural Lecture Maxwell had said that the aim of an experiment of research was 'to obtain a numerical estimate of some magnitude'. Of course, this is only part of research. Just as important is to discover the existence of new effects which no-one had previously thought of—as Faraday, Maxwell's idol had done. In Maxwell's period of tenure, very little of this sort of experiment was done at the Cavendish. Yet under J. J. Thomson, who says he ran the laboratory on similar lines to Maxwell, an enormous amount of 'pure' research was done. What was the difference between the two regimes?

In a later passage in his Inaugural Lecture, Maxwell set out his ambitions for the laboratory

> Our principal work, however, in the Laboratory must be to acquaint ourselves with all kinds of scientific methods, to compare them, and to estimate their value. It will, I think, be a result worthy of our University, and more likely to be accomplished here than in any private laboratory, if, by the free and full discussion of the relative value of different scientific procedures, we succeed in forming a school of scientific criticism, and in assisting the development of the doctrine of method.[21]

Not very inspiring, but Maxwell was under some pressure. The Cavendish was being sniped at by detractors, and it was important for Maxwell to establish its worth and scientific respectability fairly rapidly. In the mocking verse he wrote, or may have written, about himself (p.106) there is an obvious reference to the critical fire to which he was being subjected; a reviewer in Nature, with characteristic insight, expressed a pious hope that the Cavendish might creep to the level of a provincial German university in 10 years.[22]

That must have stung, and Maxwell's school of scientific criticism has a very German ring to it, an attempt to earn for the Cavendish the academic respectability of Germanic scientific method.

Another factor was that experiments which push into uncharted territory are very risky for the student involved. If something turns up, then all is well, but if nothing emerges then the student must appear to have frittered away his time. Within the fixed tenure of a scholarship or fellowship it was fairly important that a student did at least have something to show for it by the end. Precision measurements are a safe bet—some results must be obtained.

Maxwell had already been involved in a couple of long-shot experiments. In the 1860s he attempted to detect the motion of the earth relative to the aether, but with no success. Again, the first student at the Cavendish attempted to detect electromagnetic waves, but without any success. After that, it seems, he put the students onto bankers. (Not, of course, that precision measurements may not also show up interesting new phenomena.)[23]

There was also a feeling then current that physics had almost run its course. With the unification of electricity, magnetism and light, physicists could just about explain the whole of the natural order: all that was left was to measure things to a few more significant figures. No less a figure than Kirchhoff was a firm believer in this notion, as Schuster found on a visit to Heidelberg in 1872/3.

Maxwell refused to accept this, and thought one route to discovery lay precisely in accurate measurement: from the Inaugural Lecture again

> This characteristic of modern experiments—that they consist principally of measurements,—is so prominent, that the opinion seems to have got abroad, that in a few years all the great physical constants will have been approximately estimated, and that the only occupation which will then be left to men of science will be to carry on these measurements to another place of decimals.

If this is really the state of things to which we are approaching, our Laboratory may perhaps become celebrated as a place of conscientious labour and consummate skill, but it will be out of place in the University, and ought rather to be classed with the other great workshops of our country, where equal ability is directed to more useful ends.

But we have no right to think thus of the unsearchable riches of creation, or of the untried fertility of those fresh minds into which these riches will continue to be poured. It may possibly be true that, in some of those fields of discovery which lie open to such rough observations as can be made without artificial methods, the great explorers of former times have appropriated most of what is valuable, and that the gleanings which remain are sought after, rather for their abstruseness, than for their intrinsic worth. But the history of science shows that even during that phase of her progress in which she devotes herself to improving the accuracy of the numerical measurement of quantities with which she has long been familiar, she is preparing the materials for the subjugation of new regions, which would have remained unknown if she had been contented with the rough methods of her early pioneers. I might bring forward instances gathered from every branch of science, showing how the labour of careful measurement has been rewarded by the discovery of new fields of research, and by the development of new scientific ideas.[24]

In the final analysis, Maxwell was not really an experimental physicist—he was a theoretician. He had the theoretician's feel for symmetry, completeness and neatness, but not the experimenter's feel for orders of magnitude, numbers, for big and small effects.

Neither was Maxwell by nature really practically-minded. If called upon to design a laboratory he could do so, but left to itself, his mind did not stay anchored to earthly things. In 1878 he delivered the Rede lecture at Cambridge, choosing as his topic Bell's recent invention, the telephone. It was an exceedingly elegant discourse:

When, about two years ago, news came from the other side of the Atlantic that a method had been invented of transmitting, by means of electricity, the articulate sounds of the human voice, so as to be heard hundreds of miles away from the speaker, those of us who had reason to believe that the report had some foundation in fact, began to exercise our imaginations in picturing some triumph of constructive skill . . .

When at last this little instrument appeared, consisting, as it does, of parts, everyone of which is familiar to us, and capable of being put together by an amateur, the disappointment arising from its humble appearance was only partially relieved on finding that it was really able to talk.

But perhaps the telephone, though simple in respect of its material and construction, may involve some recondite physical principle, the study of which might worthily occupy an hour's time of an academic audience: I can only say that I have not yet met anyone acquainted with the first elements of electricity who has experienced the slightest difficulty in understanding the physical process involved in the action of the telephone. I may even go further, and say that I have never seen a printed article on the subject, even in the columns of a newspaper, which showed a sufficient amount of misapprehension to make it worth preserving—a proof that among scientific subjects the telephone possesses a very exceptional degree of lucidity. . . .

> This perfect symmetry of the whole apparatus—the wire in the middle, the two telephones at the end of the wire, and the two gossips at the ends of the telephones—may be very fascinating to a mere mathematician, but it would not satisfy an evolutionist of the Spencerian type, who would consider anything with both ends alike to be an organism of a very low type, which must have its functions differentiated before any satisfactory integration can take place.[25]

etc. But the lecture missed the real point, the telephone was not a toy, on which to hang witty poetic allusions, but an invention which was very shortly to change society.

Admittedly, it was a hard step to imagine that the device he was attempting to demonstrate, whose main problem was that it was so feeble, that it was almost impossible to get Garnett far enough away that the direct sound of his voice did not drown out the whole performance, would one day festoon the country with copper cables. Rayleigh said at the time

> Yesterday I had an opportunity of seeing the telephone which everybody has been talking about. The extraordinary part of it is its simplicity. A good workman might make the whole thing in an hour or two. I held conversation with Mr. Preece from the top to the bottom of the house with it and it is certainly a wonderful instrument, though I suppose not likely to come much into practical use.[26]

Edison would not have agreed, nor probably Thomson, whose fortune was made from the transatlantic telegraph.

The only times that Maxwell became interested in practical problems were when they concerned Thomson. In 1857 he suggested in a letter to Thomson the use of kites to be attached to transatlantic telegraph cables as they were being paid out off the back of a ship. The kites, pulled through the water would provide 'lift' and thereby stretch the cable properly as it sank to the ocean floor. Failure to do this had apparently caused unnecessary wastage of cable.[27] Then again in 1874, Thomson presented a paper to the I.E.E. on deep-sea sounding which was a great problem to submarine-cable engineers. He proposed using piano wire, and had found a way to splice together separate lengths of cable

> The correctness of the theory was confirmed at the Meeting by Clerk Maxwell, who asked "whether cross-filing (roughening) of the wires at a splice for increasing the grip would do so much harm by weakening the wires as to counterbalance the advantage of the increased hold." This, of course, was Maxwell's polite way of explaining that it would. His friend replied that marine glue had been found to answer so well that roughening was unnecessary.[28]

It is interesting to speculate what might have happened if Thomson had not stayed in Glasgow, where it 'was convenient to get mechanical work done'. Perhaps he would have been able to change the situation that the reformers were so worried about at the time, the divorce of British industry (partly because of education) from science. Significantly in the *other* Cambridge, M.I.T. was founded in 1865. Maxwell did not want his laboratory to have anything to do with the 'great workshops of our country'.

Maxwell was an establishment figure, though perhaps a slightly impish one; he was elected as a Conservative member to the Council of the Senate of the

University, was President of the Cambridge Philosophical Society, and a member of Eranus, a group of ex-Apostles, now successful bishops, professors etc. who continued to meet to read their philosophical essays. Maxwell's essays are published in Campbell's biography, and make an interesting comparison with his undergraduate essays: no longer full of the taut, artifical epigrams of youth, but the clear, original, deep and well-written product of a mature mind.

The opening of the Cavendish has sometimes been regarded as working a revolution in science education, but that is not really true. Laboratory teaching was insinuated slowly and with hardly a ripple into the Cambridge system. Maxwell was too much the traditionalist and conservative to want to disrupt the situation: he believed in steady change. Part II of the Natural Science Tripos was created in 1882—but practically no-one took it, because few *schools* were teaching science. That was the real problem, and the Cavendish did not touch on it. What Maxwell did do was lay the foundation of the quite unparalleled Cambridge research record.

For Maxwell himself, this period was not so productive, scientifically. Lecturing takes up a good deal of time, the amount of preparatory work is undiminished even if only a few students attend. His lecturing style had not improved, as H.F. Newall, later to become professor of Astrophysics at Cambridge, noted

> The lectures were very attractive, and I still remember being immensely struck with the contrast between the careful precision with which Maswell chose his words in defining terms, and the somewhat rambling remarks that he made in explaining the use of them. He spoke quite informally as if conversing with a friend, and sometimes even as if speaking to himself. Every now and then a humorous remark would fall from him to the obvious bewilderment of some in the small audience, but much to the unrestrained amusement of Garnett, who nearly always attended the lectures, sitting in a chair on Maxwell's side of the lecture table.[29]

His audiences were often pitifully small, only two attended his last lecture course in 1879, though this was as much to do with the Tripos system as Maxwell's qualities as a lecturer. Undergraduates were still too busy with 'paying work' for their coaches to bother with extraneous subjects like physics.

> The bulk of Cambridge undergraduates working for degrees attended college lectures and "coached" with private tutors, but the University Professors, even those of world-wide fame, such as Stokes, Adams, Maxwell or Cayley, had very small classes. Maxwell's lectures were rarely attended by more than half-a-dozen students, but for those who could follow his original and often paradoxical mode of presenting truths, his teaching was a rare intellectual treat, a lifelong inspiration, and a treasured memory.[30]

Maxwell's main occupation, though, was the editing of the Cavendish papers. Henry Cavendish had been an exception to the established English rule: he was an English nobleman who was fascinated by science. He was also a recluse who lived in his laboratory and had his meals passed in to him through a hatch so as not to disturb his work. He published only two electrical papers in his career—one being about an artificial electric eel he had constructed to prove it was possible to get an electric shock in water—but left 20 packets of unassorted papers behind when he died. These had passed into the hands of the Duke of Devonshire who expressed a desire that his ancestor's works be organized and published.

Maxwell could easily have avoided this onerous task and many have since wished he had done so, devoting more of his Cambridge time to his own research. But Maxwell was always fascinated by the history of science, and the way it worked.

> It is true that the history of science is very different from the science of history. We are not studying or attempting to study the working of those blind forces which, we are told, are operating on crowds of obscure people, shaking principalities and powers, and compelling reasonable men to bring events to pass in an order laid down by philosophers.
>
> The men whose names are found in the history of science are not mere hypothetical constituents of a crowd, to be reasoned upon only in masses. We recognise them as men like ourselves, and their actions and thoughts, being more free from the influence of passion, and recorded more accurately than those of other men, are all the better materials for the study of the calmer parts of human nature.
>
> But the history of science is not restricted to the enumeration of successful investigations. It has to tell of unsuccessful inquiries, and to explain why some of the ablest men have failed to find the key of knowledge, and how the reputation of others has only given a firmer footing to the errors into which they fell.
>
> The history of the development, whether normal or abnormal, of ideas is of all subjects that in which we, as thinking men, take the deepest interest.[31]

It is an interesting sidelight on nineteenth century physicists that there was so much interest in the history of science. Just as Maxwell 'rediscovered' Cavendish, so Rayleigh rediscovered Waterston, and Kelvin, Sadi Carnot the French engineer who was the first to formulate a version of the second law of thermodynamics. Maxwell also dug up Roger Cotes' work on optics (see p.169) and furthermore, while at the Cavendish planned to write a history of dynamics; a draft synopsis gives a list of authors with whom Maxwell was familiar: Wren, Wallis, Huyghens Hooke, Newton, Cotes, Smith, Attwood, Whewell, David Gregory, J. Playfair, Ivory, Leibnitz, D'Alembert, Euler, the Bernoullis, Laplace, and Lagrange. . . .[32] It is interesting that in the testimonial he wrote for Maxwell for the Professorship at Aberdeen, William Thomson specifically mentions Maxwell's interest in the history of science as being a recommendation

> He is possessed of one very available and important qualification for a University Lecturer,—in his extensive and familiar knowledge, not only of modern discovery, but of historical works in science: and another, even more important for clear and accurate teaching, though far less frequently met with, he possesses to a very remarkable degree, in his power of appreciating the value and meaning of a physical hypothesis.[33]

Perhaps the world of science was then so small and its internal communications so haphazard, that the chances of finding some result useful to a current research project buried in an obscure journal were high, but equally, I am sure, all these scientists were interested in the mechanics of discovery, on which the history of science shed illumination.

Maxwell's ability as a historian of science is perhaps best shown by this passage

> The cultivation and popularization of correct dynamical ideas since the time of Galileo and Newton have effected an immense change in the language and ideas of common life, but it is only within recent times, and in consequence of the increasing importance of machinery, that the ideas of force, energy and power have become accurately distinguished from each other. Very few, however, even of scientific men, are careful to observe these distinctions; hence we often hear of the force of a cannon-ball when either its energy or its momentum is meant, and of the force of an electrified body when the quantity of its electrification is meant.[34]

This almost Marxist analysis of mechanics in the nineteenth century is absolutely spot on.

In this particular case, Maxwell had been interested in Cavendish even before coming to Cambridge. In a cryptic note to Thomson, written October 15 1864, he remarked

> Do you think of doing the Cavendish expt? I have been some time devising a plan for doing it in a vacuum tube like a T upside down in a cellar in the country.[35]

It is difficult to work out which Cavendish experiment he is referring to.

Even so the vast bulk of the Cavendish papers must have given him pause, 'I am just going to walk the plank with them' he wrote to Thomson at the outset of the work.[36] But once he had begun, he soon found them fascinating. Cavendish had quietly anticipated many of the important results of the following century, he had performed some extraordinarily accurate experiments with the crudest of equipment, using two pith balls, on strings, which repelled each other to measure charge, and his own body to measure resistance. Maxwell was also fascinated by the character of the man. Maxwell in the introduction to *The Electrical Researches of the Honourable Henry Cavendish* wrote

> Cavendish cared more for investigation than for publication. He would undertake the most laborious researches in order to clear up a difficulty which no one but himself could appreciate, or was even aware of, and we cannot doubt that the result of his enquiries, when successful, gave him a certain degree of satisfaction. But it did not excite in him that desire to communicate the discovery to others which, in the case of ordinary men of science, generally ensures the publication of their results.[37]

The tidying up of Cavendish's papers was an act of posthumous respect to a very great scientist; the massive volume was finally published only a few weeks before Maxwell's own death.

In going through the papers, Maxwell found many of the experiments mentioned so original that they seemed worth repeating, checking or improving. Cavendish had devised a neat test of the inverse square law of electrostatic repulsion: if it was true, a charged conductor touched to the inside of a conducting sphere should give up all its charge. If the inverse square law did not hold, some charge should remain. Cavendish had tested this with his pith balls, and shown that if the electorstatic law was really $F \sim r^{-(2+q)}$, then the magnitude of q could not be greater than 1/60. Maxwell and MacAlister repeated the experiment with the improved technology then available, and got a maximum value for q to be 1/21,600. Another precision result, yielding an integral answer; it must have pleased Maxwell.

He also discovered that Cavendish had performed some remarkable measurements of electrical resistances, finding that distilled water becomes a better conductor after being kept standing for a year than when fresh, measuring the way that the conductivity of brine varies with the salt concentration, and, most amazing of all, Maxwell found that Cavendish had discovered Ohm's law 50 years before Ohm.

Cavendish had done his experiments by making himself part of the electrical circuit, and noting how intense were the electrical shocks he felt under different circumstances. To check his conclusions he would summon his servant Richard to replace him, and then record his reactions.

With the precision measurements of 'Ohm's law' proceeding in the same laboratory, Maxwell decided to repeat Cavendish's experiments to see just how reliable they were. 'Every man his own galvanometer', he said. Replacing the servant, Richard, all the students and visitors to the laboratory were volunteered into taking part. Maxwell discovered that Cambridge rowing men had a high resistance—due no doubt to the calloused state of their hands, and he mortally offended a touchy young American physicist, Samuel Pierpoint Langley (inventor of the bolometer) who arrived to find the great Maxwell with his sleeves rolled up, and his hands in two basins of water. Naively assuming everyone would be as interested in Cavendish's results as he was, Maxwell tried to persuade Langley to become a guinea-pig. Langley refused and left in a huff 'When an English man of science comes to the United States we do not treat him like that'.[39]

Interestingly, Ohm's law had another prediscoverer, John Leslie, who later became the Professor of Physics at Edinburgh University. In 1791 he timed how fast a Leyden jar would discharge through a piece of paper smeared in charcoal. Then he repeated the experiment cutting the paper in half, lengthwise and breadthwise. The results were a beautiful proof of Ohm's law, and rather more reliable than Cavendish's but the Royal Society of Edinburgh refused to publish the paper!

As well as the Cavendish papers, tending to his wife took up a lot of Maxwell's time. She was permanently in poor health, and Campbell records that at one point Maxwell did not sleep in a bed for three successive weeks, lecturing and running the laboratory during the day, and sitting by his wife all night. He transcribed all the Cavendish papers into legible form, longhand, by himself, by candlelight in the long nights sitting at her bedside.

Marking proofs for his books took up a lot of his time too, drawing exasperated remarks on the costiveness of publishers, and that their universal motto seemed to be 'A stitch in nine saves time'.

Of his own collected papers, 58 out of the 101 were published in this Cambridge period, but of these, 12 are book reviews, 6 are lectures, 8 are articles he wrote for the Encyclopaedia Britannica, 9 are review papers and 21 are short (though often important) notes. Only in 1878 (when his other commitments began to ease off?) does Maxwell seem to have returned to full production, with the two powerful, late papers on gas theory. Then his fatal illness began.

In his illness he lost none of his courage or humour. Tait and another Edinburgh educated colleage, Balfour Stewart had produced a book in 1875 called *The Unseen Universe*, a curious mixture of science popularization, theology and pure speculation. It was an attack on the clockwork, predictable universe of Newton. Laplace had pointed out that if you knew the positions and velocities of all the particles at any one time, then you could predict the entirety of future history, according to Newtonian mechanics. There was no room in such a universe for free will, or, seemingly, God. Tait and Stewart appealed to the aether, of course, the great unseen element of the Universe, in which God's thought could richochet and echo around, affecting the material world we inhabit, and in the aether, too, perhaps our own thoughts could be preserved, giving us an aetherial afterlife.

The public, of course, lapped it up; it rushed through 7 reprints before Tait and Stewart brought out a sequel *The Paradoxical Philosophy* in 1878. Maxwell would have none of it, he greeted the news of the new arrival with a pun on a popular music-hall song:

> It is said in *Nature* that *U.U.* is germinating into some higher form. If you think of extending the collection of hymns given in the original work, do not forget to insert 'How happy I could be with Ether'.[39]

In the review he wrote for *Nature* when the book did appear, Maxwell was harsh on the attempt to justify the aetherial afterlife by scientific jargon:

> No new discoveries can make the argument against the personal existence of man after death any stronger than it has appeared to be ever since men began to die, and no language can express it more forcibly than the words of the Psalmist: —
> "His breath goeth forth, he returneth to his earth; in that very day his thoughts perish."[40]

Here Maxwell introduces his own analogy of the soul as a train driver (p.124): he does not like the mechanistic universe any more than Tait and Balfour, but Paradoxical Philosophy was no answer

> Personality is often spoken of as if it were another name for the continuity of consciousness as reproduced in memory, but it is impossible to deal with personality as if it were something objective that we could reason about. My knowledge that I am is quite independent of my recollection that I was, and also of my belief that, for a certain number of years, I have never ceased to be. But as soon as we plunge into the abysmal depths of personality we get beyond the limits of science, for all science, and, indeed, every form of human speech, is about objects capable of being known by the speaker and the hearer. Whenever we present to talk about the Subject we are really dealing with an Object under a

false name, for the first proposition about the Subject, namely, "I am," cannot be used in the same sense by any two of us, and therefore can never become part of science at all.

The progress of science, therefore, so far as we have been able to follow it, has added nothing of importance to what has always been known about the physical consequences of death, but has rather tended to deepen the distinction between the visible part, which perishes before our eyes, and that which we are ourselves, and to shew that this personality, with respect to its nature as well as to its destiny, lies quite beyond the range of science.[41]

The review was taken in good humour and Maxwell was able to continue the joke in his last letter to Headstone (Tête, Peter) on August 28 1879

Headstone in Search of a New Sensation
While meditating, as is my wont on a Saturday afternoon, on the enjoyments and employments which might serve to occupy one or two of the aeonian aetherical phases of existence to which I am looking forward, I began to be painfully conscious of the essentially finite variety of the sensations which can be elicited by the combined action of a finite number of nerves, whether these nerves are of protoplasmic or eschatoplasmic structure. When all the changes have been rung in the triple bob major of experience, must the same chime be repeated with intolerable iteration through the dreary eternities of paradoxical existence? The horror of a somewhat similar consideration had as I well knew driven the late J.S. Mill to the very verge of despair till he discovered a remedy for his woes in the perusal of Wordsworth's Poems . . .

I have been so seedy that I could not read anything however profound without going to sleep over it. $\frac{dp}{dt}$ [42]

The last line is a reference to the progress of his illness.

In his last letter to Stokes, that same week, he even ventured to crack a joke

My wife got caught in the rain on the 13th, and on Monday I was afraid of bronchitis, but the doctor thinks that it has taken a better turn now, but she is very much distressed with neuralgia in the face and with toothache. She was much pleased by getting a letter from Mrs. Stokes this morning.

Among the subscribers to Dr. Smith's Optics, 1738, appears the name of "Mr. Gabriel Stocks." I dare say however your optical studies were already somewhat advanced (from a heredity point of view) in 1738.[43]

Maxwell died on November 5, 1879 of cancer of the abdomen. Campbell had noted there was 'some failure of the old superabundant energy' from the beginning of the year; in the Easter term he had been able to give his lectures and no more. During the summer he returned to Glenlair, improved initially, but soon regressed. He knew it was now just a matter of months, and, in great pain, returned to Cambridge to have his trusted doctor close by, to be with friends, and, as Rev. Sturrock surmises, so that Mrs. Maxwell would not have to move house immediately afterwards. He was in permanent pain but remained cheerful to the last.

At the end of his biography, Campell present some tributes to Maxwell. Here are two extracts from people who knew him best, firstly Professor Hort

Perhaps the most noteworthy of Maxwell's characteristics was his absolute independence of mind, an independence unsullied by conceit or consciousness. Preserved by his simplicity and humility from any fondness for barren paradox, he endeavoured always to see things with his own eyes, without regard to the points of view assumed on one side or another in ordinary controversy.[44]

the other from Campbell himself

> The leading note of Maxwell's character is a grand simplicity. But in attempting to analyse it we find a complex of qualities which exist separately in smaller men. Extraordinary gentleness is combined with keen penetration, wonderful activity with a no less wonderful repose, personal humility and modesty with intellectual scorn. His deep reserve in common intercourse was commensurate with the fulness of his occasional outpourings to those he loved, his respect for the actual order of the world and for the wisdom of the past, was at least as steadfast as his faith in progress. While fearless in speculation, he was strongly conservative in practice.

> In his intellectual faculties there was also a balance of powers which are often opposed. His imagination was in the highest sense concrete, grasping the actual reality, and not only the relations of things. No one was ever more impatient of mere abstractions. Yet few have had so firm a hold upon ideas. Once more, while he was continually striving to reduce to greater definiteness men's conceptions of leading physical laws, he seemed habitually to live in a sort of mystical communion with the infinite.[45]

Campbell had visited Maxwell at Glenlair during the summer of 1879, and mentions that Maxwell had been keen to show him once again the family treasures: the water colours of his childhood, his paper on ovals, his toys. The house party went down to the Orr to look at the scene of his tubbing exploits—the longest walk Maxwell was then capable of.

The reverence for continuity and tradition was one aspect of the polarity of which Campbell speaks, for it was combined with a brilliant scientific originality. Maxwell often managed to combine these two aspects of his character. He had made various 'hedra' at school, he used them to demonstrate geometry in his lectures later on. As a boy he became very skilful with the diabolo (the toy that looks like an eggcup on a skipping rope) as a scientist he made the dynamical top, and used it for his experiments on colour vision. As a boy he played with a toy called the 'magic disc' or phenakistoscope, a rotating drum, viewed through a hole in the side. By spinning it, the pictures on a paper cartridge inside would appear to move—an early cartoon film. As a scientist, he improved the device enormously by adding a viewing lens which froze the lateral motion of the cartoons, he rechristened it the 'zoetrope' and used it to demonstrate the collision of vortex rings, and the shapes of different curves by drawing his tadpoles wriggling to the different patterns. As a boy he electroplated beetles and played marbles ('bools') as a man he unified electromagnetism and founded statistical mechanics. There is a wholeness in the circularity and immutability of his interests.

CHAPTER 14

Today Maxwell is generally regarded as representing the high water mark of classical physics. With his model of gases, Newtonian mechanics extended its domain to the atom, and with his aether, it took over electromagnetism and light. Yet at the time of his death, Maxwell was probably regarded by most of his contemporaries as a brilliant eccentric, as undoubtedly a man of dazzling imagination, but was he sound?

Since Fresnel's work in 1819, physicists had been searching for a mechanical model for the luminiferous aether—the medium in which light travels. Considerable progress had indeed been made; the reflexion and refraction coefficients had been guessed at by Fresnel, and Thomson had invented a feasible if not particularly plausible mechanical aether to account for the results. Maxwell's electromagnetic *field* equations seemed to by-pass the aether, he might profess still to believe in an aether, but it was no longer necessary to him to explain light. His was a new and utterly disturbing *sort* of theory, in which he talked of waves, but did not care what was doing the waving. What Maxwell had done was introduce abstract, formal mathematical modelling by differential equations, as the basis of a physical understanding of the phenomena, rather than, as had happened with previous wave equations in physics, the mathematics dropping out of a mechanical model of the medium after suitable approximations and simplifications. It is not surprising perhaps that Thomson was perturbed by this.

> I never satisfy myself until I can make a mechanical model of a thing. If I can make a mechanical model, I understand it. As long as I cannot make a mechanical model all the way through I cannot understand, and that is why I cannot get the electromagnetic theory of light. I believe firmly in an electromagnetic theory of light, and that when we understand electricity and magnetism and light we shall see them all together as part of a whole. But I want to understand light as well as I can without introducing things that we understand even less of. That is why I take plain Dynamics. If I can get a model in plain Dynamics, I cannot in Electromagnetism. [1]

and Thomson continued to hope that he would be able to derive the electromagnetic equations from the particular internal workings of his own 'froth' model of aether. Apart from the stunning numerical connection between the speed of light and the speed of his electromagnetic waves, Maxwell's theory had little to support it at the time of his death. His prediction of the connection between dielectric constants and refractive indices had been checked for a few materials, and sometimes it worked and sometimes it did not. Maxwell's explanation was reasonable enough

> We can hardly expect even an approximate verification when we have to compare the results of our sluggish electrical experiments with the alternations of light, which take place billions of times in a second.[2]

but a useful theory must not have too many loopholes through which all its substance (i.e. its predictions) may simply drain away.

Maxwell and his students spent a lot of time at the Cavendish doing experiments on dielectric constants to check the theory, but the much more crucial experiment, trying to generate and detect electromagnetic radiation outside the known wavelengths for light, seems to have been almost completely ignored. The closest anyone came to it was the very first student to arrive at the Cavendish, William Hicks, who attempted to measure directly a speed for electromagnetism in 1874.

The principle behind the experiment was exactly the same as Faraday had used in the eighteen-fifties, and about which he had written to Maxwell (see p 143). A magnetic needle was suspended between two coils, a small one nearby, and large one far off, chosen to give the same field at the needle. Both coils were in the same circuit, and Hicks hoped that the effect of the small one might be felt by the needle and give it a twitch before the balancing effect of the big coil came into play. But, as Hicks ruefully said, 'Of course nothing came of it'.[3]

This was not really a test for the existence of electromagnetic radiation, but rather a test to see if the well-established effect of electromagnetic induction was *not* instantaneous. It involved trying to measure time down to microscopic fractions of a second and was very difficult.

To look for a new sort of radiation was even worse: a new radiation would of necessity fail to affect any existing detectors (such as the eyeball), and so how could one detect it? If it was electromagnetic radiation, it would have to consist of oscillating electric and magnetic fields, and these would be detectable, but their effect would have to be clearly distinguished from the oscillating electric and magnetic fields that are produced by ordinary induction. Perhaps it is not so surprising that no-one in the Cavendish could think of a way of pursuing the experiment.

Actually just after Maxwell's death, electromagnetic radiation *was* detected by an American inventor, David Hughes, living in London. To produce radiation a primary oscillating electric current was the first necessity; that was easy enough, the discharge of a Leyden jar was well known to do just that (as Thomson had shown). The next problem was to detect the radiation that emanated. Hughes had invented a suitable receiver, a carbon rectifying junction which he connected up to the earpiece of an ordinary telephone. A distinct noise was heard when the primary circuit was going, and if the primary was connected to a spark gap, the frying noise became deafening, just like today's telephones. Hughes was convinced he had found some new effect, a sort of atmospheric conduction, he thought, and demonstrated it to Stokes amongst others.

> The transmitter and receiver were in different rooms, about 60ft apart. After trying successfully all distances allowed in my residence in Portland-street, my usual method was to put the transmitter in operation and walk up and down Great Portland-street with the receiver in my hand, with the telephone to the ear.[4]

It says much for Stokes' devotion to science that he, as President of the Royal Society, was prepared to walk gravely up and down a busy London street with a telephone clamped to his ear, but no wires attached to it.

Stokes failed to be convinced, however, even by this bravura performance: ordinary induction was his verdict. He was correct in dismissing atmospheric conduction , but failed completely to see in the apparatus a possible means of testing Maxwell's ideas. This shows how little Maxwell's theory had yet been accepted (if Mawell himself had still been alive, he would almost certainly have heard of the experiments, and it is difficult to think *he* would not have spotted their significance).[5]

In fact Hughes was almost there; his letter to *The Electrician* continues:

> The sounds seemed to slightly increase for a distance of 60 yards, then gradually diminish, until at 500 yards I could no longer with certainty hear the transmitted signals. What stuck me as remarkable was that, opposite certain houses, I could hear better, whilst at others the signals could hardly be percieved.

That was just the clinching evidence Heinrich Hertz was to produce for electromagnetic radiation in 1887—though he understood its significance.

Helmholtz was one of the few who took Maxwell's ideas seriously (his experiments with Boltzmann on the dielectric constant—refractive index relationship have already been mentioned, p.162), and he discussed with his student Hertz the chance of detecting Maxwell's displacement (or extra) current; he set the problem as a class prize to his students, and later that year (1879) set it as a major Berlin Academy prize in the hope of tempting Hertz to tackle it.

It took Hertz eight years to find the solution. Maxwell had himself suggested that a sensitive galvonometer 'properly constructed' might detect an oscillating displacement current in a dielectric during the discharge of a Leyden jar—but did not say how properly to construct it.

Hertz originally planned to put a big block of paraffin between the plates of a condenser, put an oscillating discharge through it, and try to detect the electromagnetic effects of the displacement current in the paraffin. To measure them, Hertz reversed Hughes' apparatus and used a spark gap as a detector: he hoped there would be sufficiently strong electric fields in the paraffin to generate sparks across a narrow gap between two wires.

To his initial horror, however, he found that his apparatus was sparking everywhere, not just in the paraffin. He then realised that what he was observing was not a displacement current in paraffin, but the propagation of electromagnetic radiation through the air. He threw away the paraffin, switched to a high-frequency spark discharge as his source, and obtained some beautiful results: he could put the source at the focus of a parabolic mirror, and whatever was produced was projected as a tightly defined beam across his laboratory, to cause secondary sparks a long distance off. Objects put in the way cast a shadow. The clinching evidence was to put a large metal sheet behind the detector to reflect back the radiation. In this way interference between the arriving and reflected wave caused standing waves, and as the distance between spark-gap detector and metal sheet was varied, sparks flashed vigorously and disappeared, as the detector passed through nodes and antinodes in the standing waves.

By this means Hertz could measure the wavelength of the radiation. By converting the radiation into standing waves, Hertz had converted Maxwell and Hick's problem of measuring tiny times into one of detecting much more appreciable distances—in this case, 9 metres between successive points where no electric sparks were observed. Any effect that came and went regularly like this simply had to be a wave. Incidentally, this explained Hughes' observation that in front of some houses his device picked up a good current and by others it picked up almost nothing.

After Hertz's triumphant work, Maxwell's views on electromagnetism rapidly convinced all but the most rabid (which, unfortunately included Thomson).

In gas theory, Maxwell had himself provided convincing experimental evidence for the truth of his ideas with his measurements of viscosity. Many other confirmatory experiments soon followed. The problem of γ however, had, never been cleared up. In the lectures delivered in Baltimore where he had attacked Maxwell's electromagnetic theory,[1] Thomson also regarded γ as one of the two clouds hanging over the clear horizons of nineteenth century physics. Boltzmann had guessed an answer (see p.124) involving internal vibrations and the aether, but Maxwell knew this had to be wrong. The more internal degrees of freedom given to a molecule (the more possible ways it could absorb energy) the worse the problem of γ became.

Maxwell's difficulty was exactly the same, in fact, as that which faced Rayleigh and Jeans and led to the ultimate solution of the problem. Tyndall had made some inaccurate measurements of the amount of heat radiated away (as opposed to being lost by convection or conduction) from a hot wire as a function of its temperature (he guessed the temperature of the wire by its colour). The German physicist, Stefan, showed that Tyndall's 'measurements' implied that the rate of heat loss went as the fourth power of the Absolute temperature (the temperature calculated from Absolute zero, $-273°C$).

On this flimsy empirical base, Boltzmann considered the radiation inside a black body (1884). The black body is a delightful invention of the theoretical physicist: it is a completely enclosed cavity, whose walls are held at a constant temperature. The thermally-generated electromagnetic radiation inside the cavity should depend entirely on this temperature, and should be completely independent of the substance used to make the cavity's walls. The problem is, of course, that if the cavity is completely enclosed, then there is no way for an observer outside to measure what is going on inside, and check the theoretician's calculations (but a small hole in the side of the cavity, through which can emerge a tiny sample of the radiation, makes no significant difference to the proceedings inside). Boltzmann applied Maxwell's electromagnetic theory to work out the energy and momentum of the radiation inside a black body, and showed that, on Maxwell's theory, Tyndall and Stefan should be correct!

Rayleigh (the Cavendish Professor after Maxwell) and James Jeans then considered the black body further, and worked out what wavelengths of electromagnetic radiation could be found inside. They then applied Maxwell's equipartition principle to each individual wavelength (considered as a separate, independent vibration of the aether)—and achieved nothing but disaster. At long wavelengths, their results matched observation quite well, but at short wavelengths, they predicted a black body should emit an infinite amount of energy. Just as Maxwell had to do with his molecules (p.125), a way had to be found to 'freeze' out these unwanted degrees of freedom. Rayleigh himself was well aware of this. He had earlier written this about Maxwell's equipartition theorem.

> We are here brought face to face with a fundamental difficulty, relating not to the theory of gases merely, but also to general dynamics. In most questions of dynamics a condition whose violation involves a large amount of potential energy may be treated as a *constraint*. It is on this principle that solids are regarded as rigid, strings as inextensible, and so. And it is upon the recognition of such constraint that Lagrange's method is founded. But the law of equal partition disregards potential energy. However great may be the energy required to alter the distance of the two atoms in a diatomic molecule, practical rigidity is never secured, and the kinetic energy of the relative motion in the line of junction is the same as if the tie were of the feeblest. The two atoms, however related, remain two atoms, and the degrees freedom remain 6 in number.
>
> What would appear to be wanted is some escape from the destructive simplicity of the general conclusion relating to partition of kinetic energy, whereby the energy of motions involving larger amounts of potential energy should be allowed to be diminished in consequence. If the argument, as above set forth after Maxwell, be valid, such escape must involve a repudiation of Maxwell's fundamental postulate as practically applicable to systems with an immense number of degrees of freedom.[6]

He could have repudiated the whole of the equipartition theorem, as Thomson did, but, Rayleigh, like Maxwell, felt that there was still something valid about it. Indeed when Thomson tried to show mathematically that equipartition was actually invalid in certain cases, Rayleigh was able to show that Thomson's proofs were themselves erroneous. So when now he found the black-body formula blow up in his face, Rayleigh saw this as a final convincing proof that something extra was needed for the equipartition theorem

> It seems to me that we must admit the failure of the law of equipartition in these extreme cases. If that is so, it is obviously of great importance to determine the reason.[7]

The first step was taken by another German physicist Wilhelm Wien. According to Albert Einstein, Wien was struck by the similarity between the shape of the frequency distribution of light emitted by a black body, as observed in experiment, and the shape of the velocity distribution function for gases as calculated by Maxwell. In particular, just as high molecular speeds were chopped off by Maxwell's expotential function, so Wien guessed there was an exponential to chop off the unwanted high-frequency radiation.

To explain how this was possible, Max Planck subsequently invented quantum theory, which has revolutionised physics in the twentieth century, utterly shattering the classical synthesis Maxwell's work had seemed to complete.

In fact, the seeds of the quantum theory were there all along. In his experiments to detect electromagnetic radiation Hertz had noticed that sparks leaped across his spark-gap better if light was shone on the metal contacts. We understood this now as the first observation of the photoelectric effect; if a quantum of light is sufficiently energetic, it can give the electron in the wire which absorbs it enough kinetic energy to escape from the wire, to jump right out of the wire, just as a rocket leaving the earth needs a certain critical velocity (the escape velocity) to leave the earth's gravitational field. Once outside the wire, the free electron can easily initiate a spark.

There had been unrecognised observations of quantum effects even before that. In the letter to Tait, where Maxwell announces he has witnessed Crookes' radiometer (p.128) he continues:

> On Thursday I saw conductivity of selenium as affected by light. It is most sudden. Effect of a copper heater insensible, that of the sun great.[8]

The electrical resistance of selenium is dramatically affected by radiation containing vibrations above a certain frequency (these are present in sunlight, but not in the low temperature radiation from the copper heater). Maxwell was so bemused by this strange behaviour that he had the experiments verified at the Cavendish. Once again this is actually a quantum effect, electrons in selenium need to absorb a quantum of light of sufficient energy to be able to move around freely before the selenium will conduct electricity.

If Maxwell's photographic experiment had not gone *wrong*, he would have observed there another quantum effect. The energy of a quantum of red light is not sufficient to break up a silver iodide molecule in the wet-collodion process (that of a blue quantum is). On the then-current theories, if the iodide was irradiated long enough by sufficiently intense red light, it should be disrupted. Quantum physics says that the intensity of the radiation (i.e. the total energy in it) is irrelevant: what counts is the energy per quantum of light—and red light will never expose the plate. By giving the plate very long exposures in red light, Maxwell was actually letting enough of the miniscule amount of (energetic) ultraviolet light present in sunlight through to expose the plate. Maxwell may have thought he was validating the accepted theory, but he was actually missing the discovery of a quantum effect!

The other great revolution in twentieth century physics has been relativity. Its harbinger was the Michelson-Morley experiment, an unsuccessful attempt to measure a velocity for the earth relative to the aether. This inexplicable result was, incidentally, Thomson's other 'dark cloud'—and this is really a considerable tribute to *his* pictorial, synoptic imagination, his ability to see the way the different areas of physics connected with each other.

The speed of a wave normally has a fixed value relative to the medium in which it is travelling, the speed of sound in air is fixed relative to air, it does not depend on the speed of whatever object is making that sound. Light as a wave in the aether should have a constant speed relative to that aether, which should then act as a backdrop to the Universe, a reference frame against which an Absolute velocity could be defined. The earth as it travels in space should have a measurable Absolute velocity relative to the aether.

The idea of an Absolute velocity goes back to Newton, who, when he set up his system of dynamics, assumed that motion took place in an Absolute frame of reference, in Absolute space and Absolute time. Newton knew these were big assumptions, for which his own equations offered no proof. Newton's laws of motion deal only with acceleration; it is easy to tell when a body is accelerating because there is then a force on it. But when a body is moving at a constant velocity, there is no way of telling what that speed is (Newton's 1st law of motion. It is common experience that when two trains are parked at adjacent platforms, it is difficult for a passenger to tell whether it is his train or the other which pulls out first, or at least it is difficult for the first few moments.) Thus Newton's own laws of motion show no evidence for his assumed Absolute frame of reference.

Most of the scientists in the following two centuries forgot the uncertainty of Newton's original assumption, and failed to draw any clear distinction between Absolute and relative velocity. For earthbound experiments, the earth was at rest, for astronomers the sun was at rest.

Maxwell *was* aware of the important distinction between relative and Absolute motion: he had been teaching caution to his students since his Aberdeen days; it was another aspect of his deeper philosophical outlook on science than that of his contemporaries. It was therefore a natural question for him to ask whether his own laws of electromagnetism shed any new light on the issue, whether they gave an Absolute frame of reference where Newton's laws did not, particularly since the fields in his equations depended directly on the motions of currents, charges and magnets. In Articles 600 and 601 of the *Treatise* he investigated this possibility and convinced himself that there was no change in the form of his equations for moving frames of reference; just as in Newton's laws, the equations were indifferent to whether the observer was moving or static. Electromagnetism said nothing about Absolute motion.

Actually Maxwell had cheated. He looked at only one of his equations, the one for electromotive force ($\mathbf{E} = -\frac{\partial \mathbf{A}}{\partial t} - \nabla\phi$) he had assumed the form of $\mathbf{A}$ in the new frame of reference, and then thrown away the $\nabla\phi$ term by considering only closed circuits (Ampère's trick, which he had himself previously criticised). He *had* to do this, since he was trying to prove Newtonian (Galilean) invariance for his equations, when in fact they are already invariant under a much deeper symmetry, they are already fully relativistically invariant, as Einstein later showed.

Maxwell must have intuitively sensed there was something very special going on here, not only because he fudged the answer, but because he kept repeating the conclusion he had thereby reached.

> It is shewn, in my treatise. . . . that the currents in any system are the same, whether the conducting system or the inducing system be in motion, provided the relative motion is the same.[9]

The clearest statement of this idea came in his last book *Matter and Motion*, published somewhat improbably by the Society for the Promotion of Christian Knowledge. The book is a short account of Newton's laws of motion and his theory of gravity, but it is remarkable for the care with which Maxwell goes into the foundations of the subject. He continually stresses the provisional nature of the idea of Absolute space and time, and the relativity of our knowledge

> Absolute space is conceived as remaining always similar to itself and immovable. The arrangement of the parts of space can no more be altered than the order of the portions of time. To conceive them to move from their places is to conceive a place to move away from itself.
>
> But as there is nothing to distinguish one portion of time from another except the different events which occur in them, so there is nothing to distinguish one part of space from another except its relation to the place of material bodies. We cannot describe the time of an event except by reference to some other event, or the place of a body except by reference to some other body. All our knowledge, both of time and place, is essentially relative . . .
>
> Our whole progress up to this point may be described as a gradual development of the doctrine of relativity of all physical phenomena. Position we must evidently acknowledge to be relative, for we cannot describe the position of a body in any terms which do not express relation. The ordinary language about motion and rest does not so completely exclude the notion of their being measured absolutely, but the reason of this is, that in our ordinary language we tacitly assume that the earth is at rest . . .
>
> There are no landmarks in space; one portion of space is exactly like every other portion, so that we cannot tell where we are. We are, as it were, on an unruffled sea, without stars, compass, soundings, wind, or tide, and we cannot tell in what direction we are going. We have no log which we can cast out to take a dead reckoning by; we may compute our rate of motion with respect to the neighbouring bodies, but we do not know how these bodies may be moving in space.[10]

Einstein himself could not have given a clearer statement of the Principle of Relativity. Maxwell's own Christian beliefs come through in only one place.

> When a man has acquired the habit of putting words together, without troubling himself to form the thoughts which ought to correspond to them, it is easy for him to frame an antithesis between this relative knowledge and a so-called absolute knowledge, and to point out our ignorance of the absolute position of a point as an instance of the limitation of our faculties. Any one, however, who will try to imagine the state of a mind conscious of knowing the absolute position of a point will ever after be content with our relative knowledge.[11]

Yet Maxwell underpinned his electromagnetic equations in the aether, and here, as in most things concerning the aether, Maxwell was ambivalent. He seems to have convinced himself, albeit erroneously, of the invariance of his equations, but then failed to realise that this invariance *must* automatically extend to the aether, whose sole purpose was to enable the electromagnetic fields to obey his equations. Instead he did seem to believe that motion relative to the aether may be detectable; that the aether did indeed provide an Absolute frame of reference.

> ... the expression "at rest" has no scientific meaning, and the expression "in motion," if it refers to relative motion, may mean anything, and if it refers to absolute motion can only refer to some medium fixed in space.[12]

There were two main theories of the aether, Fresnel's and Stokes'. Fresnel had observed that light is transversely polarised, and assumed the aether was therefore a rigid solid (since only solids have shear resistance and can transmit transverse waves). There was a problem then, in that the earth is observed to move through the aether with no resistance; crashing its way endlessly through a rigid transparent plate of aether would be expected to do *something* to the earth's motion. Fresnel assumed that ordinary matter was so 'gross' that the aether just slipped through it, like water through a seive.

Stokes assumed that aether was a viscoelastic fluid: for the high vibration frequencies of light it acted like a solid, but for the relatively slow speed of the earth it flowed, and the earth just pushed it aside as it passed through.

In 1864 Maxwell thought he had invented a test between those two aether models. On Stokes' theory the aether actually at the earth's surface is static, in Fresnel's it is moving, and in moving through a glass prism carrying light, it should affect the deflection of the light by the prism. Maxwell set up a sensitive experiment to detect this extra aetherial deflection, but could detect none, whether the light came from right to left or from left to right. There was just an outside chance that he had chosen to do the experiment on the one day in the year when the earth's motion round the sun, and the sun's motion through space combined to give a zero resultant velocity against the aether. Maxwell checked this by repeating the experiment six months later, but still obtained a null result.

When he wrote up his work and submitted it to the Royal Society, Stokes pointed out that he had made a mistake. A very similar experiment to Maxwell's had been performed by the French physicist Arago in 1818, and it was to *explain* this null result that Fresnel had introduced an extra term in his theory: a solid body drags along and compresses some aether inside itself as it moves. The drag term was *designed* to compensate exactly for that part of the deflection that depends linearly on the earth's velocity through the aether. Maxwell had got his analysis wrong. His experiment could in reality check only those terms depending on the square of the earth's velocity, which then had to be divided by the square of the speed of light, and so all such effects were miniscule.

Maxwell later broadened his analysis to show that *all* earthbased tests of Fresnel's aether would depend on v^2/c^2, and therefore could not possibly succeed. Maxwell was thus, not surprisingly rather suspicious of the claim by the French physicist, Fizeau, to have detected an aether drift (1859) in some polarisation experiments. The only real way forward, Maxwell thought, was in astronomical experiments.

The first measurement of the speed of light (1675) had been astronomical. Ole Römer, a Danish astronomer, had been puzzled that the periods between successive eclipses of the moons of Jupiter were not observed to follow the predictions of gravity—which worked so miraculously everywhere else in the solar system. Then he noticed that the fluctuations in these periods depended on whether the earth was on the far or near side of the sun from Jupiter. Römer made the deduction that the fluctuations therefore had nothing to do with gravity but all to do with the fact that light took longer to reach the earth when it had to cross the extra distance to the far side of the sun. The eclipse was always on time, it just took the observer on earth a bit longer to see it.

In 1879 David Todd, the director of the Nautical Almanac in Washington D.C. published up-to-date observations of the motions of Jupiter's moons, and sent a copy to Maxwell. Maxwell immediately realised that here was an astronomical check on motion throught the aether. If the *expected* delay in receiving light from Jupiter was calculated from the known orbits of the planets and the speed of light, then an 'aether wind' in a fixed direction, would sometimes hurry the light to the earth, and sometimes try to blow it straight back into Jupiter. Therefore inaccuracies in the calculated tables could reveal an aether wind, and Maxwell wrote to Todd. The idea was interesting but beyond the experimental accuracy then attainable.

It had an unexpected side effect, though. One of Todd's staff was an ambitious young physicist, Albert A. Michelson. He might well have been shown the original of Maxwell's letter by Todd, but he certainly did see it when Todd had it published in *Nature*.

In the letter, Maxwell not only made his suggestion about Jupiter's moons, but he also repeated his assertion that all earth-based experiments were bound to fail.

> [(astronomical observations)] afford the *only* method, so far as I know, of getting any estimate of the direction and magnitude of the velocity of the sun with respect to the luminiferous medium . . . in the terrestrial methods of determining the velocity of light, the light comes back along the same path again, so that the velocity of the earth with respect to the aether would alter the time of the double passage by a quantity depending on the square of the ratio of the earth's velocity to that of light, and this is quite too small to be observed.[13]

To say something is impossible in science is nearly always fatal, particularly when brash young men with reputations to carve are within earshot.

Michelson promptly set out to prove Maxwell wrong, and invented his famous interferometer to do so. Maxwell's mistake lay not in his analysis of the experiment, but in saying that v^2/c^2 effects are too small to be seen. The difference in the time of the 'double passage' is indeed tiny, but if it is multiplied by the speed of light—which is huge—then it becomes not quite so tiny a distance. This is exactly the same trick that Hertz had spotted and used in his experiments on electromagnetic radiation, and which Maxwell and the Cambridge team had failed to realise. Michelson performed his first aether-drift experiment that year, 1880, in Germany, and despite the vast increase in accuracy made possible by his new device, obtained a null result.

The situation now became muddled. Fizeau's results (he had done another experiment on the speed of light in moving water) supported Fresnel. Michelson's results supported Stokes' aether, for they seemed to show it must be static at the earth's surface. Then Lorentz showed that Stokes' analysis of his own theory was mathematically incorrect, and that Michelson's analysis of his experiment was wrong. He had made a mistake which made the predicted result for the Fresnel aether twice as big as it should be. If this was taken into account, then his definitive experimental observation was swallowed up by the experimental inaccuracies. Michelson repeated his experiment with improved equipment and with the assistance of Edward Morley in 1887, again finding the same null result. If Lorentz' *analysis* showed Stokes' aether to be wrong, then the Michelson-Morley *experiment* showed Fresnel's aether to be wrong also (though it is worth pointing out that, unlike Maxwell they did not repeat their experiment six months later).

Other people tried to follow up the aether-drift experiment under different conditions. Oliver Lodge set up a Michelson interferometer between two horizontal circular saw blades rotating at high speed (and horrible danger). If the earth pushed the aether aside as it moves, maybe a spinning sandwich of aether would be carried around by the saw blades as they spun. Nothing was detected.

Then Fitzgerald guessed that maybe there was an effect of the aether on the motion of light, but that the aether would have the same effect on the motion of a material body: the effect of the aether wind blowing through a material body might be to bend it, squash it or stretch it just as it did to the path of a beam of light. Lorentz independently had the same idea, but *he* turned to Maxwell's equations, and proved mathematically that if matter was held together by electromagnetic forces, then Fitzgerald was exactly right. Maxwell's equations, if treated properly for moving frames of reference, gave just this effect. Maxwell had been correct all along when he had said that his equations could say nothing about Absolute motion. Once more the aether had made itself invisible.

The argument was finally closed by Albert Einstein who effectively said 'Forget the aether, all motion is relative, but the speed of light is an Absolute constant.' From those two hypotheses he derived the Special Theory of Relativity, $E = mc^2$, etc. etc. It took twelve more years of unremitting mathematical labour for Einstein to complete his great work by formulating his General Theory of Relativity—a relativistic formulation of gravity. The key to success lay in his observation that the idea of *mass* which occurs in Newton's laws of motion is identical to the idea of mass occurring in his law of gravity—yet there seems to be no *necessary* connection between the two, no reason why accelerational mass should be linearly related to gravitational mass. Einstein took this experimental fact and built his theory of gravity around it. Of the physicists before Einstein, only Maxwell seems to have noticed and been bothered by this unnecessary identity of the two types of mass. Again it is in his 'simple' introduction to Newton's ideas, *Matter and Motion*, that Maxwell explores the point most fully. After specifying that mass is a dynamic quantity, defined by Newton's laws of motion, so that a pound mass is that quantity of matter which is given a certain acceleration when pulled by a certain spring extended a certain amount, he continues

> It is an observed fact that bodies of equal mass, placed in the same position relative to the earth, are attracted equally towards the earth, whatever they are made of; but this is not a doctrine of abstract dynamics, founded on axiomatic principles, but a fact discovered by observation, and verified by the careful experiments of Newton,* of the times of oscillation of hollow wooden balls suspended by strings of the same length, and containing gold, silver, lead, glass, sand, common salt, wood, water, and wheat.
>
> The fact, however, that in the same geographical position the weights of equal masses are equal, is so well established, that no other mode of comparing masses than that of comparing their weights is ever made use of, either in commerce or in science.
>
> *"Principia," III., Prop.6.   [14]

and after properly giving the credit for this observation to Newton rather than Galileo, Maxwell went on to write

> This is the most remarkable fact about the attraction of gravitation, that at the same distance it acts equally on equal masses of substances of all kinds.[15]

Maxwell noticed the paradox here, but could make no progress. It was entirely Einstein's work which converted the observation into a beautiful physical theory.

The General Theory of Relativity, and quantum mechanics, the theory invented in the nineteen twenties where the idea of the quantum was given a mathematical framework, and relativistic quantum field theory, where the ideas of quantum mechanics and relativity were combined into a consistent mathematical synthesis, have all been couched in the language of local differential field equations. The use of this mathematical language has been the hallmark of twentieth century physics. It was the revolutionary step of introducing this sort of mathematics into physics—which so upset the mechanically-minded Thomson (p.196)—that Einstein thought was, in the final analysis, the greatest of Maxwell's achievements.

> It was Maxwell who fully comprehended the significance of the field concept; he made the fundamental discovery that the laws of electrodynamics found their natural expression in the differential equations for the electric and magnetic fields.[16]

This led to the electromagnetic theory of light, 'one of the greatest triumphs in the grand attempt to find a unity in physics'.

> The precise formulation of the time-space laws of those fields was the work of Maxwell. Imagine his feelings when the differential equations he had formulated proved to him that electromagnetic fields spread in the form of polarized waves and with the speed of light! To few men in the world has such an experience been vouchsafed.[17]

REFERENCES

Chapter One

1 C.A. Coulson in *Clerk Maxwell and Modern Science* ed. C. Domb, London (1963) p.43/44.
2. *The Life of James Clerk Maxwell*, by Lewis Campbell and William Garnett, London (1882) pp.367-9.
   From now on we will refer to this invaluable source as C. & G. There was a second edition in 1884, which contained some extra letters, of great interest, and a reprint published by the Johnson Reprint Corporation, New York (1969) which keeps the pagination of the 1st edition and includes these extra letters on Roman-numbered pages. Except where so stated we will refer to the 1st edition.
3. *The Collected Papers of James Clerk Maxwell*, ed. W.D. Niven, Cambridge (1890) Vol.2, p.358.
   There will also be frequent references to this book, which from now on will be referred to simply as the Papers.
4. *The Art of the Soluble*, P. Medawar, London (1969) p.97.

Chapter 2

1. *The Importance of Being Ernest*, Oscar Wilde, Act 1.
2. Letter from Joseph Black to Princess Dashkova (1787), EULMS Gen 837/111/36-39, quoted in James Hutton's *Theory of the Earth: The Lost Drawings*, ed. G.T. Craig, Edinburgh (1978) p 3/4/5.
3. *ibid.*, Preface. Quote taken from F. Ellenberger
4. *ibid.*, Chapter 2, pp 6-10
5. *ibid.* p.25
6. *Outlines of a Plan for Combining Machinery with the Manual Printing Press*, Edinburgh Medical and Philosophical Journal, Vol X, 1831
7. C. & G. p.185.

Chapter Three

1. *Corsock Parish Church, its Rise and Progress*, the Rev. Geo. Sturrock, Castle Douglas (1899) p.17.
2. *Thomas Carlyle, A History of the First Forty Years of his Life*, J.A. Froude, London (1882) Vol.2 p.420.
3. C. & G. p.257
4. Papers Vol.2 p.751

5. C. & G. p.262
6. *Sir Donald Macalister of Tarbet*, by Edith Macalister, London (1935) p.48
7. These stories come from C. & G., p.28, 31.
8. C. & G. p.27
9. C. & G. p.29.
10. C. & G. p.29.
11. C. & G. p.43.
12. C. & G. p.34
13. C. & G. p.33

Chapter Four

1. *The Edinburgh Academy Register*, 1824-1914, p.106-112.
2. Plan of the Edinburgh Academy, 1847, p.3.
3. *The Chronicles of the Cumming Club*, Lt. Colonel A. Fergusson, published privately, Edinburgh (1887) p.39.
4. *ibid.* p.46
5. C. & G. p.85/6.
6. *The Chronicles of the Cumming Club* p.59 & 58.
7. P.G. Tait's Obituary of Maxwell in Proc. R. Soc.Edin., Vol.10 (1878-80) p.332.
8. *The Chronicles of the Cumming Club*, p.59.
9. *ibid.* p.25/6.
10. C. & G., p.57. A trump was a Jew's harp.
11. *ibid.*, p.57/8
12. *ibid.*, p.67
13. *ibid.* p.67/8
14. *ibid.*, p.176
15. P.G. Tait's obituary of Maxwell in Proc. R. Soc. Edin., Vol.10 (1878-80) p.334
16. C. & G. p.60
17. *ibid.*, p.74-6.
18. *ibid.*, p.78.
19. *ibid.*, p.69.
20. *ibid.*, p.81.

Chapter 5

1. C. & G., p.127
2. *ibid.*, p.126
3. *Scottish Philosophy and British Physics*, R. Olson, Princeton (1975) p.19.
4. *ibid.*, p.37
5. *ibid.*, p.44
6. *ibid.*, p.118
7. *ibid.*, p.96
8. Forbes had a very Scottish pronunciation, and, it would appear, spelling. Maxwell (C. & G., p.124) once tried to capture it in a letter to Campbell: "But to make an *abrupt transcision*, as Forbes says . . ." to which Campbell appends the footnote, "Forbes was extra-precise in articulation. Hence the spelling.". Everitt notes that Forbes later tried to "correct" his pronunciation by taking elocution lessons with the celebrated actress Mrs. Siddons.
9. *Scottish Philosophy and British Physics*, p.226.
10. *ibid., p.238*

11. *The Democratic Intellect—Scotland and her Universities in the Nineteenth Century,* Edinburgh (1961) p.184.
12. *Scottish Philosophy and British Physics,* p.127.
13. C. & G., p.119
14. *Lord Kelvin and the Age of the Earth,* J.D. Burchfield, London (1975), p.93.
15. Papers Vol.1 p.155/6.
16. Papers Vol.2 p.220.
17. C. & G., p.119.
18. *ibid.,* p.268.
19. *ibid.,* p.115/6
20. *Guthrie's Physics,* Nature, Feb.6, 1879, p.311.
21. *The Democratic Intellect,* p.173.
22. *Scottish Philosophy and the British Physics,* p.22.
23. *ibid.,* p.165
24. *ibid.,* p.189
25. *ibid.,* p.190.
26. *Elementary Treatise on Electricity.* J.C. Maxwell, Oxford, 2nd edition (1888) p.46/7.
27. C. & G. p.175
28. *ibid.,* p.113, 133/4.
29. *ibid.,* p.81
30. *ibid.,* p.120-123
31. *ibid.,* p.137/8
32. *Clerk Maxwell and Modern Science,* ed. C. Domb, London (1963) p.44.
33. C. & G., p.135
34. *ibid.,* p.489

Chapter 6

1. *Five Years in an English University,* C.A. Bristed, London (3rd edition, 1873) p.290.
2. *The Life of John William Strutt, Third Baron Rayleigh,* R.J. Strutt, (Madison) (1968) ((Original edition with the author's comments added)) p.241.
3. *ibid.,* p.34.
4. *Five Years in an English University,* p.286/7.
5. C. & G., p.158.
6. *The Life of John William Strutt,* p.27.
7. *The Democratic Intellect,* G.E. Davie, Edinburgh (1961) p.117.
8. Proc. R. Soc. Edin., 1879-80, p.332.
9. C. & G., p.175.
10. *ibid.,* p.175
11. *ibid.,* p.133
12. *ibid.,* p.175.
13. *ibid.,* p.169
14. *ibid.,* p.196
15. *ibid.,* p.155
16. *ibid.,* p.417
17. *ibid.,* p.422
18. *ibid.,* p.215
19. *ibid.,* p.236/7
20. *ibid.,* p.243

21. Maxwell's inaugural lecture at King's College London, published in Am.J. Phys., vol.47, No.11, Nov. 1979, p.930.
22. Papers, Vo. 2. p.305.
23. *Ten British Mathematicians*, A. MacFarlane, New York (1916) p.107.
24. *The Life of John William Strutt*, p.39.
25. C. & G., p.217
26. *ibid.*, p.220
27. *ibid.*, p.158
28. *Michael Faraday*, L. Pearce Williams, London (1965) p.262.
29. C. & G., p.213.
30. *ibid.*, p.257.

Chapter 7

1. C. & G., p.251
2. *ibid.*, p.256
3. *ibid.*, p.281
4. *ibid.*, p.419
5. *ibid.*, p.421
6. *ibid.*, p.405
7. *ibid.*, p.253
8. *ibid.*, p.291
9. *ibid.*, p.267/8
10. *ibid.*, p.261
11. Testimonial written by Airy for Maxwell for the Chair of Natural Philosophy at the University of Edinburgh, 2/12/1859.
12. C. & G., p.264
13. *ibid.*, p.262
14. *ibid.*, p.297
15. *Aberdeen University Review*, Vol.III. No.9 June 1916. p.204.
16. *The Life of John William Strutt, Third Baron Rayleigh*, by R. J. Strutt, Madison, Wisconsin (1968) p.407.
17. Margaret Tait (grand daughter of P.G. Tait), letter to Prof. D.J. de Solla Price, Cambridge Univ., Maxwell Archives.
18. C. & G. p.318
19. *Life and Scientific Work of P. G. Tait*, by C. G. Knott., Cambridge, (1911) p.16
20. Margaret Tait to Prof. D.J. de Solla Price, Cambridge University, Maxwell Archives.
21. Letter of 18/12/1856, published in Proc. Camb. Phil. Soc., Vol.32, (1936) p.722.
22. Testimonial from John MacRobin M.D., Prof. of Medicine Marischal College and University of Aberdeen, for Maxwell for the Edinburgh Chair, written 7/12/1859.
23. Testimonial from the Rev. John Crombie, Minister of the U. P. Church and lecturer in Botany at King's College and University of Aberdeen, written 3/12/59.
24. C. & G., p.173
25. *ibid.*, p.259
26. *ibid.*, p.428/9
27. *James Clerk Maxwell*, C.W.F. Everitt, New York (1965) p.53/4.

Chapter 8

1. published in Am.J.Phys., Vol.47, No.11, Nov.1979, p.930.
2. *ibid.*, p.931
3. *ibid.*, p.931/2.
4. in *Clerk Maxwell and Modern Physics*, ed. C. Domb, London (1963) p.18.
5. C. & G., p.316
6. *Life on John William Strutt*, p.115.
7. *On the Educational System of Prussia*, Sir A. Grant, in Proc. R. Soc. Edin. Vol.7, No. 83, 1870/1 p.324.
8. *The letters of John Richard Green*, ed. L. Stephen, London (1901) p.44/5. There are many variations extant on the exact wording of the exchange between Huxley and Wilberforce. Huxley later reckoned that Green's was the most accurate, though he doubted he ever used the word 'equivocal'.
9. *Life of Charles Darwin*, F. Darwin, London (1908) p.202/3.
10. *Lord Kelvin and the Age of the Earth*, J. D. Burchfield, London (1975) p.84.
11. *The Organisation of Science in England*, D.S.L. Cardwell, London (1957), p.80, 105.
12. Published by Prof. R.V. Jones in Notes and Records of the Royal Society of London, Vol.28 (1973) p.57.
13. C. & G., p.200
14. *ibid.*, p.179
15. Papers, Vol.2, p.375/6.
16. *Lord Kelvin and the Age of the Earth*, p.32
17. C. & G., p.390.
18. *ibid.*, p.386.
19. *ibid.*, p.83
20. *Corsock Parish Church: Its Rise and Progress etc.*, Rev. George Sturrock, Castle Douglas (1899) p.20.
21. *Mathematical Gazette*, Vol.20 (1936) p.27
22. *Life of John Willian Strutt*, p.360.
23. *Michael Faraday*, L. P. Williams, London (1965) p.380.
24. *Experimental Researches in Electricity*, London (1855) Vol.III, p.447.
25. Sir Charles Wheatstone, B. Bowers, London (1975) p.22.
26. C. & G., p.385.
27. *ibid.*, p.186

Chapter 9

1. C. & G., p.420
2. *ibid.*, p.252
3. *ibid.*, p.344/5
4. *Memoir and Scientific Correspondence of Sir G.G. Stokes*, ed. J. Larmor, Cambridge (1907) p.38. We will refer to this as the Stokes Letters.
5. *The Origins of Clerk Maxwell's Electric Ideas*, as described in Familiar Letters to W. Thomson, ed. by Sir J. Larmor, Proc. Camb. Phil. Soc., vol.32, 1936, p.697. We will refer to this as the Kelvin Letters.
6. *ibid.*, p.745
7. C. & G., p.384
8. Postcard from Maxwell to Tait, in Cambridge Univesity Maxwell Archive. Published in Life and Scientific Works of P.G. Tait, C.G. Knott, Cambridge

(1911), p.100. We will refer to this as the Life of Tait.
9. *History of Physics*, F. Cajori, London (1899) p.252.
10. Life of Tait, p.177
11. Proc. R. Soc. Edin., vol.6 (1866-69) p.94.
12. Papers, Vol.2, p.445
13. Life of Tait, p.106
14. *ibid.*, p.196/7
15. *The Life of William Thomson, Baron Kelvin of Largs*, S.P. Thompson, London (1910) p.612.
16. *Ten British Mathematicians*, A. MacFarlane, New York, (1916), p.39
17. *ibid.*, p.44
18. Life of Tait p.185
19. *ibid.*, p.101
20. *The Electrician*, Vol.2, 1879, p.271
21. Life of Tait, p.143
22. Tait Archive, National Library of Scotland
23. Life of Tait, p.151.
24. *ibid.*, p.153/4
25. *ibid.*, p.99
26. *ibid.*, p.195
27. C. & G., 2nd ed. p.261, Johnson Reprint XXIII.
28. *ibid.*, p.646
29. *ibid.*, p.639-641
30. Note by John Blackwood in the Blackwood Magazine files.
31. Letter written by Maxwell 28/9/1874, Cambridge University, Maxwell Archive.
32. Life of Tait, p.175
33. *ibid.*, p.175
34. Papers, vol.2, p.782
35. *Nature*, Feb.1879, p.384. The Review appeared on p.311.
36. Papers, vol.2, p.660/1
37. C. & G., p.357
38. Notes and Records of the Royal Society of London, Vol.35 (1980). p.95-99.

Chapter 10

1. C. & G., p.85
2. Papers, Vol.1, p.291
3. C. & G., p.295
4. *The Aim and Structure of Physical Theory*, P. Duhem, tr. P. Wiener, Princeton (1954), p.71.
5. Letter from Maxwell to Kelvin, 14/11/1857, Glasgow University Archive.
6. as above.
7. Letter from Maxwell to Kelvin, 1/8/1857, Glasgow University archive.
8. *The Collected Papers of John James Waterston*, ed. J.S. Haldane, Edinburgh & London (1928) p.209
9. *ibid.*, p.209
10. Letter to P.G. Tait, 24/7/1873, Cambridge University Archive.
11. *Scottish Philosophy and British Physics*, R.C. Olson, Princeton (1975) p.175
12. *Collected Papers of John James Waterston*, p.214.
13. Stokes Letters, p.10.
14. Papers, Vol.1, p.377.

15. Stokes Letters, p.10
16. British Association Report 1860, Vol.28, pt.2 p.16
17. This is a reference to some lines by Tennyson:
       It is the little rift within the lute
       That by and by will make the music mute
       And ever widening silence all.
                              (The Idylls of the King, Merlin and Vivien, line 388).
18. Letter to P.G. Tait, 7/3/1865, Cambridge University Archive.
19. Letter to the Master of the Mint, 1/5/1865, Cambridge University Archive.
20. Papers, Vol.2, p.11
21. Royal Society Archives, RR 6 178
22. Papers, Vol.2, p.26
23. Royal Society Archive, RR 6 179; Papers Vol.2, p.700
24. Cambridge University Archive.
25. *Keaton*, R. Blesh, New York (1971) p.152
26. *Life of John William Strutt*, p.47.
27. Letter of 11/8/1873, Cambridge University Archive.
28. C. & G. p.332.
29. *Life of John William Strutt*, p.350
30. Papers, Vol.2, p.433
31. *ibid.*, p.428
32. *ibid.*, p.426
33. C. & G., p.325
34. Papers, Vol.2, p.713
35. *ibid.*, 219/20
36. Letter no.65, late April 1874, Cambridge University Archive
37. Royal Society Archive, RR 7 295
38. Royal Society Archive, RR 8 89
39. Letter of 15/5/1876, Peterhouse Library, Cambridge
40. Royal Society Archive RR 8 123, 15/6/1878
41. Letter of 31/10/1980, from Professor T.G. Cowling to the author.
42. Royal Society Archive, RR 8 188
43. *ibid.*, RR 8 189
44. *ibid.*, RR 8 196
45. Papers, Vol.2, p.703
46. Royal Society Archive, Reply to RR 8 196
47. quoted in *Biographical Fragments*, A. Schuster, London (1932) p.219.
48. Royal Society Archive, RR 8 102.
49. quoted in Biographical Fragments, p.225

Chapter 11

1. Kelvin letters, p 697
2. *Sir Isaac Newton's Mathematical Principles of Natural Philosophy and His Systems of the World*, tr A Motte, rev. F. Cajori, Berkeley (1946) p 547.
3. Letter of 18/11/1690, quoted in *The Aim and Structure of Physical Theory*, P. Duhem, tr. P. Wiener, Princeton (1954) p 15.
4. Letter from Descartes to Fr. Mersenne, 20/4/1646, quoted on p 15 of ref.3.
5. Papers, Vol.2 p 487.
6. *Matter and Motion*, J. Clerk Maxwell, London (1876) p 123.
7. *LIfe and Letters of Faraday*, H. Bence Jones, London (1870) Vol.2, p 353.

8. *A Treatise on Electricity and Magnetism*, J. Clerk Maxwell, Oxford, 2nd ed. (1881) Vol.1, p IX/X. We will refer to this book as the Treatise.
9. Kelvin letters, p 701.
10. *ibid.*, p 717.
11. *ibid.*, p 705.
12. *ibid.*, p 711/2.
13. C. & G., p 216.
14. Treatise, Vol.2, Art 528, p163/4.
15. *ibid.*, p.163.
16. Papers, Vol.1, p 451/2.
17. Treatise, Vol.1, p X
18. *ibid.*, Vol.2, p 147, Art 502.
19. Papers, Vol.1, p 155/6.
20. *ibid.*, p.207
21. *ibid.*, p.187
22. *ibid.*, p.187/8
23. *ibid.*, p.203
24. *ibid.*, p 195. Also p.183, 193.
25. Treatise, Vol.2, Art 529, p 165.
26. C. & G., 2nd edition, p.200/1. Not reprinted in the Johnson reprint.
27. *ibid.*, 1st edition, p.199
28. *ibid.*, 2nd ed. p.202, Johnson reprint p.XV
29. *ibid.*, p.204, Johnson reprint p. XVII
30. *ibid.*, 1st ed., p.288/9
31. *ibid.*, p.289/90
32. Letter of 14/11/1857, in Glasgow University Archive.
33. *Life of Tait*, p.222
34. 'On Physical Lines of Force', reprinted in Papers Vol.1, p.451-513
35. *ibid.*, p.468
36. Diagram taken from papers Vol.1 p.489
37. *ibid.*, p.492
38. *ibid.*, p.486. This closes Part 2 of the paper, which appeared in *Phil. Mag.* April and May 1861. Part 3 came out in the Jan. and Feb. 1982 issue.
39. *ibid.*, p.490/1
40. *ibid.*, p.492
41. *ibid.*, p.496
42. *ibid.*, p.500
43. C. & G., 2nd ed. p.244, Johnson reprint p.XXII.
44. Nature, Jan 31, 1878, p.258. Papers, Vol.2, p.662/3
45. Papers, Vol.1, p.452.
46. Letter of 15/10/1864. Glasgow University Archive.
47. C. & G., p.342.
48. Papers, Vol.1, p.526-597
49. *ibid.*, p.563
50. Letter to Tait of 23/12/1867. Cambridge University Archive. Published in *Life of Tait*, p.215.
51. Papers, Vol.2, p.783.
52. C. & G., p.330.
53. Treatise, Art 793, p.402.
54. Papers, Vol.1, p.528.
55. *ibid.*, p.564.

56. *ibid.*, p.582.
57. *Nineteenth-Century Aether Theories,* K.F. Schaffner, Oxford (1972) p.69.
58. Papers, Vol.2, p.138.
59. *ibid.*, p.136.
60. Letter to Tait of 8/5/1871, Cambridge University Archive.
61. Treatise, Vol.1, p.24.
62. *History of Physics,* F.Cajori, London (1899) p.253.
63. Treatise, Vol.2, Art. 831, p.427/8.
64. Treatise, Vol.1, Art 37, p.41.
65. *An Elementary Treatise on Electricity,* J. Clerk Maxwell, ed. W. Garnett, Oxford (2nd ed. 1888) Art 10.p.8.
66. *ibid.*, p.115/6 Art. 139
67. Kelvin Letters, p.739.
68. Treatise, Vol.2, Art. 569, p.196.
69. *ibid.*, art. 550, p.182.
70. J. Larmor, Proc. R. Soc., **81** (1908) p.XIX.
71. *Hermann von Helmholtz,* L. Koenigsberger, tr. F.A. Welby, Oxford (1906) p.288.
72. Letter to Thomson, 15/10/1864, Glasgow University Archive.
73. Stokes Letters, p.26.
74. Royal Society Archive, RR 8 89.
75. Treatise, Vol.1, Art. 255 p.346, Art. 260 p.350.
76. *ibid.*, Art. 261 p.353.
77. Papers, Vol.1, p.156.

Chapter 12

1. Papers, Vol.1 p.30
2. *The Structural Analysis of Gothic Cathedrals,* R. Mark, Scientific American, November 1972.
3. Papers, Vol.1, p.181
4. *ibid,* Vol.2, p.31
5. *ibid.*, p.379
6. *ibid.*, Vol.1, p.238, 271
7. *ibid.*, p.126
8. *ibid.*, p.445
9. C. & G., p.334
10. Papers, Vol.1, p.410
11. *ibid., p.514; Vol.2, p.161*
12. *Theory of Heat,* J. Clerk Maxwell, London (11th ed., 1894) p.169
13. *ibid.*, p.195
14. Stokes Letters, p.34
15. Letter of 12/10/74, Cambridge University Archive
16. Papers Vol.1, p.115
17. *ibid.*, Vol.2, p.96
18. *ibid.*, p.105
19. Tait to Maxwell, letter of 24/3/1875, Cambridge University Archive.
20. Papers Vol.2, p.538
21. Letter to Tait of 25/3/1875, Cambridge University Archive.
22. Lecture in the City Hall, Glasgow, 29/1/1880, published in the *Life of Tait,* p.296.
23. Letter to Thomson of 10/8/1872. Cambridge University Archive.
24. C. & G., p.330

25. Papers, Vol.2, p.160
26. *Clerk Maxwell and Modern Physics,* ed.C. Domb, London (1963) p.43/4.

Chapter 13

1. *Sir Charles Wheatstone,* B. Bowers, London (1975) p.59.
2. *Life of John William Strutt,* p.406.
3. *Ten British Mathematicians,* A. MacFarlane, New York (1916) p.140/1.
4. *The Organisation of Science in England,* D.S.L. Cardwell, London (1957) p.92.
5. *Life of Lord Kelvin,* S.P. Thompson, London (1912) p.563.
6. *Life of John William Strutt,* p.48/9.
7. C. & G., p.350.
8. *Life of John William Strutt,* p.102/3, 162, 217.
9. C. & G., p.389.
10. *ibid.,* p.383.
11. Papers, Vol.2, p.242/3.
12. *Life of John William Strutt,* p.50
13. Papers, Vol.2, p.247-9.
14. *ibid.,* p.242, 251.
15. *The History of the Cavendish Laboratory, 1871-1910,* London (1910) p.38/9.
16. *ibid.,* p.31.
17. *James Clerk Maxwell and Modern Physics,* R.T. Glazebrook, London (1896) p.78/9.
    See also p.46/47 of this book.
18. C. & G. p.392
19. *The History of the Cavendish Laboratory, 1871-1910,* p.22.
20. *Nature,* July 6 1876, p.207/8
21. Papers, Vol.2, p.250.
22. C. & G., p.392.
23. In *The Progress of Physics, 1875-1908,* A. Schuster, Cambridge (1911) p.44,
    Schuster puts another perspective on this issue:
    Surprise may reasonably be expressed that while Maxwell was surrounded at the
    Cavendish Laboratory by a number of young physicists who firmly believed in his
    electromagnetic theory, no attempt was made by them to furnish an experimental
    proof of their great master's theoretical deductions. ([Schuster is wrong here, there
    was one attempt]). The explanation lies to a great extent in Maxwell's habit of letting
    his students go their own way and find their own problems. Unless a student had asked
    him directly to suggest a problem, I doubt whether it would have occurred to him to
    give advice in the selection of a subject for investigation. Nevertheless, and I speak
    from personal knowledge, the desirability of an experimental proof of Maxwell's
    theory was realised by Cambridge men, and other British physicists, who were in
    contact with them, such as Fitzgerald. But the experimental difficulties seemed
    formidable, notably as regards the emission of electromagnetic waves of sufficient
    intensity to give a measurable effect at a distance. Had there been two equally
    probable theories in the field, I doubt not the attempt to carry out a crucial
    experiment would have been made, but we were perhaps over confident in the
    inherent truth and simplicity of Maxwell's conception.
24. Papers, Vol.2, p.244.
25. *ibid.,* p.742.
26. *Life of John William Strutt,* p.133.
27. Letter of 14/11/1857. Glasgow University Archive.
28. *History of the I.E.E.,* R. Appleyard, London (1939) p.54.

29. *History of the Cavendish Laboratory, 1871-1910*, p.106.
30. *Fifty Years of Electricity*, J.A. Fleming, London (1921), p.287.
31. Papers, Vol.2, p.251
32. C. & G., p.398.
33. Testimonial written on 23/12/1856.
34. *An Elementary Treatise on Electricity*, J. Clerk Maxwell, Oxford (2nd edition, 1888) Art. 97, p.76/77.
35. Letter to Thomson, 15/10/1864. Glasgow University Archive.
36. Letter to Thomson date illegible, Glasgow University Archive.
37. *The Electrical Researches of the Honourable Henry Cavendish F.R.S.*, ed. J. Clerk Maxwell, Cambridge (1879) p.XLV.
38. *Biographical Fragments*, A. Schuster, London (1932).
39. *Life of Tait*, p.242.
40. *Nature*, 19/12/1878, p.142. Papers, Vol.2, p.756.
41. *ibid.*, p.143
42. *Life of Tait*, p.244/5.
43. Stokes letters, p.44.
44. C. & G., p.418.
45. *ibid.*, p.425.

Chapter 14

1. *Lectures on Molecular Dynamics, and the Wave Theory of Light*, Baltimore, (1884) p.270.
2. Papers, Vol.2, p.772
3. *History of the Cavendish Laboratory, 1871-1910;* London (1910) p.19.
4. The Electrician, May 1899, p.41.
5. It is difficult now, when Maxwell's equations form the basis of undergraduate education in electromagnetism, to realise how small an impact Maxwell's work had on the physics community outside the small circle who gathered around him at Cambridge.
   Schuster *was* in that group, and he later had this to say:
   > The state of electrical science in 1870 must present itself somewhat differently to the scrutiny of a historian who derives his information from the records of published papers, and to the recollection of a student, who received his impressions from the teachers of the time. Maxwell's great paper 'A Dynamical Theory of the Electromagnetic Field', appeared in 1864, but I doubt whether the younger generation of physicists had their attemtion drawn to, or seriously arrested by it, before the publication in 1872 of the two volumes of *Electricity and Magnetism*. I believe that the first systematic course of lectures based on Maxwell's theory was given by myself at the Owens College [(in Manchester)] during the session 1875-6 (Sir Joseph Thomson was one of the three students attending the course).
   > *(The Progress of Physics 1875-1908*, A. Schuster, Cambridge (1911) p.6.)
6. *Life of John William Strutt*, p 352/3
7. *ibid.*, p.355
8. Letter of late April 1874, Cambridge University Archive. Schuster relates that he told Kirchhoff of this discovery. Kirchhoff firmly believed that the end of physics was nigh, that all that was left to be done was to measure a few constants a bit more accurately. When told of this finding, all he said was:
   > I am surprised that so curious a phenomenon should have remained undiscovered till now.

(*The Progress of Physics, 1875-1908,* A Schuster, Cambridge (1911) p 9).
9. Papers, Vol.2, p.290
10. *Matter and Motion,* J. Clerk Maxwell, London (1876) p.20, 84, 85.
11. *ibid.,* p.20
12. *ibid.,* p.42
13. *Nature,* 29/1/1880, p.314
14 *Matter and Motion,* p.40.
15. *ibid.,* p.113
16. *Ideas and Opinions,* A. Einstein, London (1973) p.344
17. *ibid.,* p.327

INDEX